# State-Sum Models of Piecewise Linear Quantum Gravity

# State-Sum Models of Piecewise Linear Quantum Gravity

## Aleksandar Miković
Lusófona University, Portugal

## Marko Vojinović
University of Belgrade, Serbia

**World Scientific**

NEW JERSEY · LONDON · SINGAPORE · BEIJING · SHANGHAI · HONG KONG · TAIPEI · CHENNAI · TOKYO

*Published by*

World Scientific Publishing Co. Pte. Ltd.

5 Toh Tuck Link, Singapore 596224

*USA office:* 27 Warren Street, Suite 401-402, Hackensack, NJ 07601

*UK office:* 57 Shelton Street, Covent Garden, London WC2H 9HE

**Library of Congress Cataloging-in-Publication Data**
Names: Miković, Aleksandar, author. | Vojinović, Marko, author.
Title: State-sum models of piecewise linear quantum gravity / Aleksandar Miković
    (Lusófona University, Portugal), Marko Vojinović (University of Belgrade, Serbia).
Description: New Jersey : World Scientific, [2023] | Includes bibliographical references.
Identifiers: LCCN 2023012301 | ISBN 9789811269318 (hardcover) |
    ISBN 9789811269325 (ebook for institutions) | ISBN 9789811269332 (ebook for individuals)
Subjects: LCSH: Quantum gravity. | Quantum gravity--Mathematical models. |
    Quantum gravity--Mathematics.
Classification: LCC QC178 .M55 2023 | DDC 530.14/3--dc23/eng/20230606
LC record available at https://lccn.loc.gov/2023012301

**British Library Cataloguing-in-Publication Data**
A catalogue record for this book is available from the British Library.

For any available supplementary material, please visit
https://www.worldscientific.com/worldscibooks/10.1142/13233#t=suppl

Printed in Singapore

# Preface

This book describes a novel approach in the study of quantum gravity (QG) state-sum models, which is based on the application of the effective action method from quantum field theory. Related to that is a study of the effect of a non-trivial path-integral (PI) measure on the PI finiteness, as well as a study on the dependence of the semiclassical expansion of the effective action on the PI measure.

Another novelty is a detailed study of the idea that the spacetime at small distances is not a smooth manifold but a piecewise linear (PL) manifold corresponding to a triangulation of a smooth manifold. This is a radical departure from the standard approach in PLQG, where the PL structure, i.e. the triangulation, is assumed to be non-physical and an auxiliary tool serving to define a QG theory on a smooth manifold. The main advantage of this paradigm shift is that finite QG path integrals can be constructed, while the semiclassical limit can be explored by using the effective action formalism. A smooth spacetime is then interpreted as an approximation to a PL manifold when the maximal edge length is small and the number of spacetime cells is large. The corresponding effective action can be then approximated by the usual QFT effective action with a cutoff, where the cutoff is determined by the average edge length in the spacetime triangulation. A further consequences of the idea that the spacetime is a PL manifold is that the cosmological constant has a continuous spectrum, and that the spectrum contains the observed value of the cosmological constant. We also describe some implications for quantum cosmology.

A description of higher gauge theory formulation of general relativity is also given, since the corresponding state-sum models do not suffer from the problems found in the spin-foam models of QG. These new state-sum models are called spin-cube models, and they are categorical generalizations of the spin-foam models, since one labels the edges, the triangles and the tetrahedra in a triangulation with representations of a 2-group, which is a categorical generalization of a group.

A major part of the book is devoted to the results obtained by the authors in the period from 2009 to 2016, and some more recent results have been also included. The book contains descriptions of the main PLQG approaches, but the emphasis is on a more detailed description of the Regge PLQG and the corresponding effective action. Our book can serve as an introductory text for a further research, so that it can be useful for young researchers, as well as for other researchers who are interested in this area.

We would like to thank John Barrett, Louis Crane, Laurent Freidel, Renate Loll, Steven Carlip, Ignatios Antoniadis and Hermann Nicolai for conversations over the years, who helped us to clarify our ideas.

Lisbon, March 2023

Aleksandar Miković and Marko Vojinović

# Contents

# Chapter 1

# Introduction

## 1.1 The standard approach to QG

The standard approach to the problem of constructing a quantum gravity (QG) theory [66, 67] can be described as the following problem. Let $M$ be a smooth 4-manifold, of topology $\Sigma \times I$, where $\Sigma$ is a 3-manifold and $I$ an interval from $\mathbb{R}$. Let $g$ be a Minkowski-signature metric on $M$ and $\varphi$ a set of matter fields on $M$. Then the goal is to find a triple $(\hat{g}, \hat{\varphi}, \hat{U}(I))$, where $\hat{g}$ and $\hat{\varphi}$ represent Hermitian operators parametrized by the points of $M$, acting in some Hilbert space $\mathcal{H}_*$, while $\hat{U}$ is a unitary evolution operator parametrized by $I$, such that the classical limit ($\hbar \to 0$) of the quantum time-evolution is equivalent to the Einstein equations.

There are 3 general ways to construct the operator triple and the Hilbert space:

    (1) canonical quantization,
    (2) path-integral quantization,
    (3) quantum field theory (QFT) formalism.

Note that there is also the formalism of the third quantization of general relativity (GR). This formalism serves in order to describe a possibility of the existence of many separate universes, as well as creation and annihilation of universes. Although such a situation can be modeled by a single 4-manifold $M$ whose boundary structure involves one or more than two disjoint spatial boundary components, we will not discuss this possibility in this book.

Also note that in mathematics one defines a topological quantum field theory (TQFT), see [5, 130], as a third quantization of a $d$-dimensional GR, or of some collections of $p$-forms on a $d$-dimensional manifold, via Atiyah's axioms which specify the properties of the corresponding path integral on a $d$-dimensional manifold with multiple boundary components. The Atiyah axioms capture the fact that the corresponding field theories do not have the local degrees of freedom (DoF), and only the global, i.e. the topological, DoF exist. Since GR in $d = 4$ has local DoF, it may seem that using the TQFT formalism is not appropriate. However, we will see in chapter 3 that the GR path integral can be understood as a modification of the path integral for a topological GR theory.

## 1.2   Canonical quantization of GR

In the canonical quantization approach, one does not construct the operator $\hat{U}$ directly, but solves instead the corresponding Schrödinger equation, which is known as the Wheeler-DeWitt (WDW) equation [41]. The canonical formulation of the Einstein-Hilbert (EH) action

$$S_{EH} = \int_M \sqrt{-\det g}\, R(g)\, d^4x \,,$$

is given by the ADM action

$$S_{ADM} = \int_{t_1}^{t_2} dt \int_\Sigma d^3x \left( p^{kl}\dot{q}_{kl} - n^k \mathcal{D}_k(p,q) - N\mathcal{H}(p,q) \right) \,,$$

where $(p, q)$ are canonically conjugate pairs, $\dot{q} = dq/dt$,

$$\mathcal{D}^k = -2\nabla_j p^{jk} \,,$$

$$\mathcal{H} = \sqrt{\det q} \left( K_{kl}K^{kl} - (K_l^l)^2 - R_3(q) \right) \,, \tag{1.1}$$

where $K$ is the extrinsic curvature of $\Sigma$ and $R_3$ is the intrinsic curvature of $\Sigma$. $K$ is given by

$$K^{kl} = \frac{1}{\sqrt{\det q}} \left( p^{kl} - \frac{2}{3}q^{kl}p_m^m \right) \,,$$

while the spacetime metric is given by

$$ds^2 = g_{\mu\nu}dx^\mu dx^\nu = (-N^2 + q_{kl}n^k n^l)dt^2 + 2q_{kl}n^l dt dx^k + q_{kl}dx^k dx^l \,.$$

In order to understand the significance of the Hamiltonian constraint (1.1), we will discuss the case of finitely many DoF. This happens when we consider the metrics of spacetimes with symmetries, i.e. the minisuperspace models. Let us consider a dynamical system with $n$ degrees of freedom $q_a$

and the corresponding canonically conjugate momenta $p^a$, such that the dynamics is given by the action

$$S = \int_{t_1}^{t_2} dt \, (p^a \dot{q}_a - NH(p,q)) \, .$$

This action has a reparametrization symmetry given by

$$\delta p = \epsilon\{H, p\} \, , \; \delta q = \epsilon\{H, q\} \, , \; \delta N = \dot{\epsilon} \, ,$$

which is the finite-dimensional analogue of the diffeomorphism symmetry of the EH action.

Let us assume that there exists a canonical transformation (CT)

$$\{(p^a, q_a) \,|\, a = 1, 2, \ldots, n\} \to \{(P_T, T), (P^\alpha, Q_\alpha) \,|\, \alpha = 1, 2, \ldots, n-1\} \, , \tag{1.2}$$

such that

$$H(p,q) = 0 \Leftrightarrow P_T - H_*(P, Q, T) = 0 \, .$$

The CT (1.2) can be found by solving $H(p,q) = 0$ for a $p^a$ such that $p_a = f_a(p', q)$. Consequently $P_T = p^a$, $P^\alpha = p^\alpha$ while $(T, Q^\alpha) = J^{-1}(p)(q_a, q_\alpha)$, where the matrix $J$ is given by

$$J_d^c(p) = \frac{\partial F^c(p)}{\partial p^d}$$

and $P^a = F^a(p)$.

Note that $H_*(P, Q, T)$ is known as the reduced phase-space (RPS) Hamiltonian, since it is a result of solving the constraint $H(p, q) = 0$. One then has an unconstrained dynamics given by the action

$$S_* = \int_{t_1}^{t_2} dt \left( P^\alpha \dot{Q}_\alpha - H_*(P, Q, t) \right) \, ,$$

where the reparametrization symmetry is fixed by the gauge choice $T(t) = t$.

The quantization in the reduced phase space then proceeds in the usual way, but the problem in the full GR case is that $H_*$ is a complicated non-local function, see [54]. Consequently, no progress has been achieved so far in the RPS quantization of GR.

Alternatively, on can quantize first and then solve the constraint, which is the Dirac quantization procedure. This amounts to promoting the momenta and coordinates $(p, q)$ into Hermitian operators acting in the Hilbert space $L_2(\mathbb{R}^n)$, also known as the kinematical Hilbert space. One then has to solve the operator version of the Hamiltonian constraint

$$\hat{H}(\hat{p}, q)\Phi(q) = 0 \, , \tag{1.3}$$

where

$$\hat{p}^a \Phi(q) = i\hbar \frac{\partial \Phi(q)}{\partial q_a} .$$

The equation (1.3) is known as the WDW equation, and the space of the solutions of the WDW equation has to be promoted into a Hilbert space $L_2(\mathbb{R}^{n-1})$, which is known as the physical Hilbert space. A WDW equation can define a unitary quantum evolution if we identify a time variable $T$, so that we can have the RPS Schrödinger equation

$$\left( i\hbar \frac{\partial}{\partial T} - \hat{H}_*(\hat{P}, Q, T) \right) \Psi(Q, T) = 0 .$$

In the full GR case, the WDW operator is constructed from the expression for the Hamiltonian constraint (1.1), and as it is well known, there is no unique expression which one can construct, due to the normal ordering ambiguities. A more serious problem is that nobody knows how to solve the corresponding WDW equation, which becomes a functional differential equation

$$\hat{\mathcal{H}}\left( \hat{p}(x), q(x) \right) \Phi[q(x)] = 0 ,$$

where $\hat{\mathcal{H}}$ is an operator associated to the Hamiltonian constraint $\mathcal{H}(p, q)$ such that

$$\hat{p}^{kl}(x)\Phi = i\hbar \frac{\delta \Phi}{\delta q_{kl}(x)} .$$

The main problem is finding solutions of the WDW equation beyond the minisuperspace approximation. A related problem is finding a time variable $T(p, q)$ such that in the gauge $T(p, q) = t$ the WDW equation becomes a Schrödinger equation

$$\left( i\hbar \frac{\partial}{\partial t} - \hat{H}_*(\hat{P}(x), Q(x), t) \right) \Psi[Q(x), t] = 0 ,$$

where

$$\hat{H}_* = \int_\Sigma d^3 x \, \hat{\mathcal{H}}_* .$$

### 1.2.1  *Canonical LQG*

The most developed example of the canonical quantization of GR is Loop Quantum Gravity (LQG), see [32] for a recent review and references. The basic idea is to change the canonical variables $(p^{jk}, q_{jk})$ by first passing to the triads $e_j^\alpha$ and their canonical momenta $\tilde{p}_\alpha^j$ via

$$q_{kj} = e_k^\alpha e_{j\alpha} , \quad p^{kj} = \tilde{p}_\alpha^{(k|} e^{\alpha|j)} , \tag{1.4}$$

where $e_\alpha^j$ is an inverse triad. Therefore

$$\mathcal{D}_k(p,q) = \tilde{\mathcal{D}}_k(\tilde{p}, e), \quad \mathcal{H}(p,q) = \tilde{\mathcal{H}}(\tilde{p}, e). \tag{1.5}$$

Since we have more DoF (triads have 9 components, while the spatial metric has 6 components), we need to introduce a new constraint, which corresponds to the gauge symmetry of triad rotations. This gives the constraint

$$G_\alpha = \varepsilon_{\alpha\beta\gamma}\, e_j^\beta\, \tilde{p}^{j\gamma}.$$

In this way we arrive to the triad canonical formulation of the EH action (introduced by Dirac)

$$S = \int dt \int_\Sigma d^3x \left( \tilde{p}_\alpha^j \dot{e}_j^\alpha - \lambda^\alpha G_\alpha - n^k \tilde{\mathcal{D}}_k - N\tilde{\mathcal{H}} \right).$$

By performing the following canonical transformations

$$(\tilde{p}_\alpha^k, e_k^\alpha) \to (\bar{p}_k^\alpha, \tilde{e}_\alpha^k) \to (A_k^\alpha, E_\alpha^k),$$

where $\tilde{e}_\alpha^k = (\det e)\, e_\alpha^k$ and

$$A_k^\alpha = \Gamma_k^\alpha(\tilde{e}) + \mathrm{i}\,\bar{p}_k^\alpha, \quad E_\alpha^k = \tilde{e}_\alpha^k, \tag{1.6}$$

where i$=\sqrt{-1}$, one obtains the Ashtekar canonical variables [4]. The complex $SO(3)$ connection $A$ is defined by

$$\Gamma_k^\alpha(\tilde{e}) = \varepsilon^{\alpha\beta\gamma}\omega_{k\,\beta\gamma}(e),$$

where $\omega$ is the torsion-free spin connection on $\Sigma$.

The introduction of a complex $SO(3)$ connection $A$ is necessary if one wants that the new constraints are all polynomial. Namely, the Dirac constraints are equivalent to

$$\tilde{G}_\alpha = \partial_k E_\alpha^k + \varepsilon_{\alpha\beta\gamma} A_j^\beta E^{j\gamma}, \quad \tilde{D}_j = F_{jk}^\alpha E_\alpha^k, \quad \tilde{\mathcal{H}} = \varepsilon_{\alpha\beta\gamma} F_{jk}^\alpha E^{j\beta} E^{k\gamma}, \tag{1.7}$$

which are known as the Ashtekar constraints.

Although the complex Ashtekar connection makes the GR constraints polynomial, which simplifies the problem of solving the WDW equation, one has to impose the reality conditions

$$(A_k^\alpha)^* + A_k^\alpha = 2\tilde{\Gamma}_k^\alpha(E), \quad (E_\alpha^k)^* = E_k^\alpha,$$

because the physical metric is real.

Since $\tilde{\Gamma}(E) = \Gamma(\tilde{e})$ is a non-polynomial function, this introduces a non-polynomiality back into the theory. Note that there is the Kodama proposal of holomorphic quantization [71], see also [85], where one avoids the imposition of the reality condition by using a holomorphic wavefunction, which is just a functional of $A$ and does not depend on $A^*$. The Kodama proposal works for the systems with a finite number of DoFs, but the field theory version has never been developed nor checked.

One can still introduce a real Ashtekar connection as

$$A_k^\alpha = \Gamma_k^\alpha(\tilde{e}) + \gamma \, \bar{p}_k^\alpha \,, \qquad E_\alpha^k = \frac{1}{\gamma} \, \tilde{e}_\alpha^k \,,$$

where $\gamma$ is the Barbero-Immirzi parameter [12, 64]. In this case the Hamiltonian constraint $\bar{\mathcal{H}}$ becomes non-polynomial, but the variables are real.

### 1.2.1.1   *Spin-network basis*

The Ashtekar connection allows one to define a Wilson loop invariant

$$W_\gamma(A) = Tr \, P \exp \left( \int_\gamma A \right) = Tr \, U_\gamma(A) \,,$$

for a closed loop $\gamma$ in $\Sigma$. Note that $W$ is the trace of the holonomy of $A$ along $\gamma$, and the value of $W$ depends on a representation of the Lie-algebra valued connection $A$. The functional $W$ is invariant under the $SO(3)$ gauge symmetry transformations, but it is not a diffeomorphism invariant. It becomes a diffeomorphism invariant for flat connections, i.e.

$$W_\gamma(A) = W_\gamma(\tilde{A}) \,,$$

where $\tilde{A}$ is related to $A$ by a diffeomorphism and

$$F_A(x) = \mathrm{d}A + A \wedge A = 0$$

for all $x \in \Sigma$.

One can generalize the concept of a Wilson loop invariant by replacing a loop $\gamma$ by a spin network $\hat{\gamma}$. A spin network is a closed oriented graph $\gamma$ embedded in $\Sigma$ such that we associate an $SU(2)$ irreducible representation[1] (irrep), i.e. a spin $j \in \mathbb{N}/2$, to each edge $l$ of the graph $\gamma$ and an $SU(2)$ intertwiner $\iota$ to each vertex $v$ of the graph, such that

$$\iota \in Hom \left( j_1 \otimes \cdots \otimes j_k, j_{k+1} \otimes \cdots \otimes j_{k+l} \right) \,,$$

where $j_1, j_2, \ldots, j_k$ are the spins of the incoming edges in a vertex and $j_{k+1}, \ldots, j_{k+l}$ are the spins of the outgoing edges. One can then define a gauge-invariant functional

$$W_{\hat{\gamma}}(A) = Tr \left( \prod_{l \in \gamma} U_l(A) \prod_{v \in \gamma} \iota_v \right) \,,$$

---

[1]The replacement of $SO(3)$ by $SU(2)$ is equivalent to demanding that the spinor irreps of $SO(3)$ should be included.

which is a generalization of the Wilson loop invariant and it will be diffeo-morphism invariant for flat connections.

Let $\Psi(A)$ be a gauge-invariant and diffeomorphism-invariant functional of a real connection $A$. One can heuristically define a spin-network basis by using the spin-network invariants

$$I(\hat{\gamma}) = \int \mathcal{D}A \, W_{\hat{\gamma}}(A) \, \Psi(A) \, . \tag{1.8}$$

This formula can be derived from

$$|\Psi\rangle = \int \mathcal{D}A \, |A\rangle\langle A|\Psi\rangle = \int \mathcal{D}A \, \Psi(A)|A\rangle$$

and

$$|\Psi\rangle = \sum_{\hat{\gamma}} |\hat{\gamma}\rangle\langle\hat{\gamma}|\Psi\rangle = \sum_{\hat{\gamma}} I(\hat{\gamma})|\hat{\gamma}\rangle \, , \tag{1.9}$$

where $\hat{\gamma} = (\gamma, j_\gamma, \iota_\gamma)$. Consequently the formula (1.8) is equivalent to

$$\langle\hat{\gamma}|\Psi\rangle = \int \mathcal{D}A \, \langle\hat{\gamma}|A\rangle\langle A|\Psi\rangle \, . \tag{1.10}$$

Note that by definition

$$W_{\hat{\gamma}}(A) \equiv \langle A|\hat{\gamma}\rangle = (\langle\hat{\gamma}|A\rangle)^* \, ,$$

so that in order to use the formula (1.8) the connection $A$ has to be real-valued.

The invariant (1.8) can be rigorously defined in some special cases, for example when

$$\Psi(A) = \exp\left(i\kappa \int_\Sigma Tr\left(A \wedge \mathrm{d}A + \frac{2}{3}A \wedge A \wedge A\right)\right) \, , \tag{1.11}$$

where $\kappa$ is a constant, see [80].

The relationship (1.9) can serve to define the spin-network (SN) basis. The basis states $|\hat{\gamma}\rangle$ are orthogonal and span the Hilbert space of gauge invariant and diffeomorphism invariant states $\mathcal{H}_{GD}$. This is achieved by considering abstract spin-network graphs $\gamma$, which are the homotopy classes of spin-network graphs embedded in $\Sigma$.

One can also define formally the inverse loop transform

$$\Psi(A) = \sum_{\hat{\gamma}} I(\hat{\gamma})W_{\hat{\gamma}}(A) \, ,$$

so that one can work in the SN basis and try to solve the Hamiltonian constraint. The physical Hilbert space $\mathcal{H}_{phys}$ will be a subspace of the spin-network Hilbert space $\mathcal{H}_{GD}$. In this approach, one has to construct the Hamiltonian constraint operator in the SN basis, which will be given by a matrix $H_{\hat{\gamma}\hat{\gamma}'}$. Then one has to find a solution of the WDW equation

$$\sum_{\hat{\gamma}'} H_{\hat{\gamma}\hat{\gamma}'} \, c_{\hat{\gamma}'} = 0 \,,$$

so that a physical state will be given as

$$|\Psi_{phys}\rangle = \sum_{\hat{\gamma}} c_{\hat{\gamma}} \, |\hat{\gamma}\rangle \,. \tag{1.12}$$

So far nobody has constructed an exact non-trivial solution of the type (1.12), although the Kodama wavefunction [70], which is given by (1.11) for $\kappa = i/\lambda\hbar$, and the $\delta(F)$ wavefunction from [85], could be the examples of such solutions, provided that the holomorphic quantization procedure is valid for field theories.

In the case of the Euclidean GR, the Ashtekar connection is real, so that the Kodama wavefunction and the $\delta(F)$ wavefunction are the exact solutions of the corresponding WDW equations. In the Kodama wavefunction case, the coefficients $c_{\hat{\gamma}}$ correspond to the Witten-Reshetikhin-Turaev invariants for the quantum group $U_q(su(2))$ where $q = \exp(i\pi/k + 2)$, $k = 1/(l_P^2\lambda) \in \mathbb{N}$ and $\lambda$ is the cosmological constant, see [80]. For $k \notin \mathbb{N}$, one obtains the same formula for $c_{\hat{\gamma}}$ as in the $k \in \mathbb{N}$ case, see [81].

## 1.3 Path-integral quantization of GR

The difficulties in solving the WDW equation in canonical QG can be partially resolved in the path-integral quantization. One uses the following relation

$$\langle q_2, \phi_2 \,| \, \hat{U}(t_1, t_2) \,| \, q_1, \phi_1 \rangle = \int \mathcal{D}g \, \mathcal{D}\varphi \exp\left(iS(g,\varphi)/\hbar\right) \,, \tag{1.13}$$

where $q_i$ and $\phi_i$ are the metric and the matter fields on the 3-manifold $\Sigma$ at a time $t_i$, $i = 1, 2$, while $|q, \phi\rangle$ is an eigenstate of $\hat{q}$ and $\hat{\phi}$ operators. The classical action $S(g, \varphi)$ is given by

$$S(g, \varphi) = S_{EH}(g) + S_m(g, \varphi) \,, \tag{1.14}$$

where $S_{EH}$ is the Einstein-Hilbert action, while $S_m$ is the matter action coupled to gravity.

The main problem is to define the path integral in (1.13), and the most progress has been achieved so far by using the spin foam models, see section 3.2. The idea of the SF models is to use in the formula (1.13) the spin-network basis $|\hat{\gamma}\rangle$ instead of the spatial metric basis $|q(x)\rangle$.

Consequently a spin foam $\hat{\Gamma}$, which is a colored 2-complex, is embedded in $\Sigma \times I$ such that a spin network $\hat{\gamma}_1$ is the boundary of $\hat{\Gamma}$ at $t = t_1$ and a spin network $\hat{\gamma}_2$ is a boundary of $\hat{\Gamma}$ at $t = t_2$. One can think of a spin foam as a surface (2-complex) $\Gamma$ formed by the time evolution of a spin network graph $\gamma$, so that a face $f$ of $\Gamma$ will be labeled by an irrep of the Lorentz group $\Lambda_f$, while an edge $l$ of $\Gamma$ will be labeled by an intertwiner $\iota_l$ from the *Hom* space of the irreps for the corresponding faces. Therefore $\hat{\Gamma} = (\Gamma, \Lambda_\Gamma, \iota_\Gamma)$, where

$$\Lambda_\Gamma = \{ \Lambda_f \,|\, f \in \Gamma \}, \quad \iota_\Gamma = \{ \iota_l \,|\, l \in \Gamma \}.$$

Then one can write

$$\langle \hat{\gamma}_2 \,|\, \hat{U}(t_1, t_2) \,|\, \hat{\gamma}_1 \rangle = \sum_{\hat{\Gamma}} c_{\hat{\Gamma}} \, A_{\hat{\Gamma}}(\hat{\gamma}_1, \hat{\gamma}_2), \tag{1.15}$$

where $A_{\hat{\Gamma}}$ is the spin foam amplitude for a spin foam $\hat{\Gamma}$ and $c_{\hat{\Gamma}}$ are some coefficients to be determined. The least problematic part in this formula is the SF amplitude $A_{\hat{\Gamma}}$, which is known, see section 3.3. The problematic part is the determination of the infinite sum over the spin foams. Namely, it is not known what would be the choice of the coefficients which would give a convergent sum and the correct classical limit.

Note that in (1.15) we considered the case when the matter fields $\varphi$ are absent. The presence of the matter fields modifies the boundary spin networks and the corresponding spin foam, see [84].

## 1.4  QFT for GR

The QFT approach to QG is based on the idea that for weak gravitational fields, when $|R(g)| \ll 1/l_P^2$, one can use

$$g_{\mu\nu} = \eta_{\mu\nu} + \kappa h_{\mu\nu},$$

where $\kappa^2 = G_N$ and $h$ is a spin-2 field propagating on a flat spacetime $M$ whose metric is $\eta$. Consequently the EH action becomes

$$S_{EH} = S_2(h) + \sum_{n=3}^{\infty} \kappa^{n-2} S_n(h), \tag{1.16}$$

where

$$S_n(h) = \int_M \langle \partial h \, \partial h \, h^{n-2}, \, \eta^{n+2} \rangle \, d^4 x,$$

and $\langle \partial h \, \partial h \, h^{n-2} \, , \, \eta^{n+2} \rangle$ represents a linear combination of scalars formed by contracting a product of $n$ tensors $h_{\mu\nu}$ and 2 operators $\partial_\mu$ with $n + 2$ tensors $\eta^{\mu\nu}$.

As is well known, the action (1.16) is perturbatively non-renormalizable (NR), so that the corresponding physics cannot be described by a finite number of coupling constants, and therefore the predictivity is lost. There have been proposed various strategies in order to solve this problem, and supersymmetry is the most famous proposal [48]. The idea is to introduce a symmetry which relates bosons and fermions, since in loop diagrams bosons contribute an opposite sign to the contribution of fermions. Then it may happen that if the right supersymmetric action is chosen, the loop infinities may cancel, so that the perturbation theory will be finite. In the case of the $N = 8$ supergravity action, the theory is finite up to 2 loops, but at higher loops the infinities may appear [18]. A generalization of the supergravity idea is the superstring theory, and it has been shown by Mandelstam [76] that the superstring perturbation theory is finite. However, the price for this is the introduction of an infinite tower of massive fields, as well as many new massless fields, which are not seen in the current high-energy elementary particle scattering experiments.

In the asymptotic safety (AS) approach, see [26,43], one follows the QFT approach, but instead of using the standard perturbative approach, which leads to the problem of non-renormalizability and infinitely many coupling constants, one introduces the hypothesis of the existence of a certain type of a fixed point under the renormalization group flow. Namely, at such a fixed point there are only a finite number of relevant coupling constants so that the physics will be determined by this finite set of couplings. The main problem with the AS approach is proving the existence of such a fixed point. This hypothesis could be proved or disproved if one starts with the full effective action, containing infinitely many coupling constants, and check the corresponding renormalization group equations. However, one can test the AS hypothesis for only a small number of coupling constants, i.e. one truncates the effective action to a small number of terms.

We have mentioned the superstring theory as one of the ways to resolve the NR problem of quantum GR. Basically, string theory introduces an infinite tower of massive fields, so that the action (1.16) gets modified, and becomes an action for infinitely many massive fields coupled to $h$. However, there is a simplification in string theory, since the transition amplitudes for such a QFT can be expressed by using path integrals for two-dimensional manifolds, representing the string world-sheets. Since the (super)string amplitudes are finite, there is no need for a renormalization procedure. Although string theory can be considered as a perturbatively finite theory of quantum gravity, the main problem is the necessity of using the flat background metric $h$, rendering the analysis of quantum effects in regions of strong gravitational fields very difficult. Another problem is that superstring theory can only be formulated in the anti de Sitter metric background, which corresponds to a negative cosmological constant, while we know from astrophysical observations that the cosmological constant is positive.

## 1.5 QG and short-distance structure of spacetime

Note that in the standard QG approach, the structure of the spacetime manifold $M$ is not changed after the quantization and it is well known that this is the main source of the difficulties for a quantization of gravity [66,67]. The problems of constructing a QG theory based on a smooth manifold, as well as a plausible assumption that the short-distance structure of spacetime may not be a smooth manifold, lead to an alternative approach where $M$ is replaced by some quantum spacetime $\widehat{M}$. The obvious choice would be a non-commutative manifold based on $M$, like in the case of non-commutative geometry (NCG) [33,74], where the coordinates of $M$ become elements of a non-commutative algebra.

Another choice is made in the superstring theory [56], where the coordinates of $M$ become coordinates of the loop manifold $\mathcal{L}M$ and new Grassmann (anticommuting) coordinates are added, so that $\widehat{M}$ is a loop super manifold. The appearance of the superstring theory in the 1980's as a theory of quantum gravity [56], introduced the idea that the finiteness of a quantum gravity (QG) theory can be achieved by changing the mathematical structure which describes spacetime. In the superstring theory case, this was realized by using a loop manifold instead of a manifold.

Each of these approaches has its advantages and disadvantages, and the main problem of the superstring theory is the apparent absence of supersymmetry in the elementary particle experiments, while in the case of the non-commutative geometry approach, it is a non-geometric nature of the non-commutative manifolds, so that it is difficult to see what is their physical meaning.

There is another candidate for $\widehat{M}$, which was introduced in the seminal work of Tullio Regge [111]. This is the idea of using a manifold triangulation $T(M)$ and the corresponding edge lengths to define the path integral for general relativity (GR). The quantum Regge calculus was further developed by other researchers, see [112] for a review and references. However, in their work $T(M)$ was considered only as an auxiliary structure, and the main goal was to find the smooth-manifold limit of the Regge path integral. This turned out to be an elusive task, and Regge's approach was abandoned by most of the researchers, although in the past 20 years there has been a revival of the quantum Regge calculus in the form of the spin-foam models [104] and in the form of the casual dynamical triangulations (CDT) models [3].

Recently a new idea was proposed in the context of the quantum Regge calculus, and this is to use $T(M)$ as the short-distance structure of spacetime [88, 100]. Hence the basic structure of spacetime is a piecewise linear (PL) manifold $T(M)$, while the smooth manifold $M$ is considered as an approximation valid when the number of spacetime cells (4-simplices) is large and their size is small. Therefore $\widehat{M} = T(M)$, and this idea is analogous to a situation in hydrodynamics, where a fluid, which is a large collection of molecules, can be approximated as a continuous medium at the scales much larger than the intermolecular distance.

## 1.6    QG and piecewise-flat manifolds

As we pointed out, Regge was the first to use $T(M)$ in order to define the path integral for general relativity (GR) [111], see [59] for a modern review. However, in Regge's approach the triangulation was an auxiliary structure and had to be removed via the smooth limit $T(M) \to M$. However, obtaining the smooth limit in the Regge approach is a difficult problem. The same applies to the case of spin-foam models of LQG, which can be only defined when spacetime is a PL manifold. In causal dynamical triangulations (CDT) approach [3], $T(M)$ is also used to define the path integral, but it is also considered an auxiliary structure. Obtaining the smooth limit in CDT is proposed by performing a sum over the triangulations.

Note that in the spin foam (SF) approach to QG, the construction of the SF amplitudes can be done only if one uses a PL manifold $T(M)$. Because of the PL context, the GR path integral becomes a state sum, i.e. the sum over the labels associated to the simplices of the simplicial complex, and one sums the corresponding amplitudes. When those labels are continuous, the sums are replaced by the integrals. By an appropriate choice of the amplitude, the state sum (SS) can be made finite, see [97], but what is less easier to see is what is the classical limit of the state sum. The easiest way to see what is the classical limit of a SS model is to use the effective action approach [94]. Determination of the smooth-manifold limit for a state-sum model is a more difficult and still unsolved problem.

A simple way to circumvent the smooth-manifold limit problem is to assume that $\widehat{M} = T(M)$, see [88,99]. We will refer to this approach as the piecewise linear quantum gravity (PLQG) approach, and as we mentioned in the previous section, the main idea is to assume that the short-distance structure of spacetime is not given by a smooth manifold $M$, but by the PL manifold $T(M)$. This means that a triangulation of $M$ becomes physical, i.e. an intrinsic property of spacetime, instead of just being an auxiliary structure which serves to define the smooth-manifold limit. This new paradigm can be justified by the fact that when $M$ is a compact manifold, then it is difficult to distinguish between $M$ and $T(M)$ when the number of simplices $N$ in $T(M)$ is large. For example, a picture of a circle is hard to distinguish from a picture of a regular polygon with a large number of the edges. As far as the dynamics of a PLQG model is concerned, one can use the quantum effective action, and then it is not difficult to see that for large $N$, the effective action for $T(M)$ is very well approximated by the QFT effective action for $M$, where one uses the QFT with a cutoff determined by the average edge length in the triangulation. This is analogous to the fluid dynamics case, where a fluid can be approximated by a continuous medium on the length scales which are much bigger than the intra-molecular distance, but still small enough compared to the dimensions of the bodies immersed in the fluid. Consequently, instead of solving a dynamical problem for a large number of molecules, one uses the Navier-Stokes partial differential equations for the density and velocity fields of a fluid.

The choice $\widehat{M} = T(M)$ is much simpler than the one made in NCG or in the superstring theory. Also, by using $T(M)$ one reduces the infinite number of the degrees of freedom (DoF) for $g$ and $\varphi$ to a finite number, which then simplifies the quantization.

# Chapter 2

# Classical theories of gravity on PL manifolds

## 2.1 Regge formulation of general relativity

The Regge discretization of GR [111], amounts to replacing the smooth spacetime manifold $M$ with a simplicial complex $T(M)$ which corresponds to a triangulation of $M$, while the metric on $T(M)$ is determined by the set of the edge lengths

$$\{L_\epsilon > 0 \mid \epsilon \in T(M)\}. \tag{2.1}$$

Given the edge lengths (2.1), one would like to define a metric on the PL manifold $T(M)$ such that the PL metric on each 4-simplex $\sigma$ of $T(M)$ is flat and of the desired signature. The physical signature is $(-, +, +, +)$, or equivalently $(+, -, -, -)$, but Regge used the Euclidean signature $(+, +, +, +)$, because it was considered a simpler case.

The PL metric can be constructed by using the Cayley-Menger (CM) metric [59]

$$G_{\mu\nu}(\sigma) = L_{0\mu}^2 + L_{0\nu}^2 - L_{\mu\nu}^2, \tag{2.2}$$

where the five vertices of $\sigma$ are labeled as $0, 1, 2, 3, 4$ and $\mu, \nu = 1, 2, 3, 4$. Although the CM metric is flat in a 4-simplex, it is not dimensionless and hence it is not diffeomorphic to $g_{\mu\nu} = \delta_{\mu\nu}$. This can be remedied by defining a new PL metric

$$g_{\mu\nu}(\sigma) = \frac{G_{\mu\nu}(\sigma)}{2L_{0\mu}L_{0\nu}}, \tag{2.3}$$

which is a rescaled CM metric such that $g_{\mu\nu}$ is dimensionless. The scaling factor is chosen such that $g_{\mu\nu}$ is related to $\delta_{\mu\nu}$ by linear homogeneous coordinate transformations such that the coefficients are dimensionless.

Ensuring the Euclidean signature of a PL metric requires the following restrictions

$$\det G(\sigma) > 0\,, \tag{2.4}$$

$$\det G(\tau) > 0\,, \tag{2.5}$$

$$\det G(\Delta) > 0\,, \tag{2.6}$$

for every 4-simplex $\sigma$, every tetrahedron $\tau$ and every triangle $\Delta$ of $T(M)$. The last inequality is equivalent to the triangular inequalities for the edge lengths of a triangle. These inequalities permit us to define the volumes of $n$-simplices via the Cayley-Menger determinants [30]

$$\det G(\sigma_n) = 2^n (n!)^2 V^2(\sigma_n)\,, \quad n = 2, 3, 4\,. \tag{2.7}$$

Note that for an arbitrary assignment of $L_\epsilon$, the volumes $V_n$ can be positive, zero or imaginary. Just taking the strict triangular inequalities will ensure the positivity of the triangle areas, but then some of the higher volumes can be zero or negative. Hence all of the three inequalities must be imposed.

The Einstein-Hilbert (EH) action on $M$ is given by

$$S_{EH} = \int_M \sqrt{\det g}\, R(g)\, d^4x\,, \tag{2.8}$$

where $R(g)$ is the scalar curvature associated to a metric $g$. On $T(M)$ the EH action becomes the Regge action

$$S_R(L) = \sum_{\Delta \in T(M)} A_\Delta(L)\, \delta_\Delta(L)\,, \tag{2.9}$$

where $A_\Delta = V(\Delta)$ is a triangle area and the deficit angle $\delta_\Delta$ is given by

$$\delta_\Delta = 2\pi - \sum_{\sigma \supset \Delta} \theta_\Delta^{(\sigma)}\,, \tag{2.10}$$

where $\theta_\Delta^{(\sigma)}$ is a dihedral angle. It is defined as the angle between the 4-vector normals associated to the two tetrahedrons that share the triangle $\Delta$, and it is given by

$$\sin\theta_\Delta^{(\sigma)} = \frac{4\,A_\Delta V_\sigma}{3\,V_\tau V_{\tau'}}\,. \tag{2.11}$$

### 2.1.1 *Minkowski PL metric*

The novelty in the Minkowski case is that $L_\epsilon^2$ can be positive, negative or zero, so that $L_\epsilon \in \mathbb{R}_+$ or $L_\epsilon \in i\mathbb{R}_+$ or $L_\epsilon = 0$. Consequently we have to indicate in $T(M)$ which edges are space-like (S), time-like (T) or light-like (L). Although one can triangulate a pseudo-Riemannian manifold such that all the edges are space-like, it is much simpler and more natural to use the triangulations where we have both the space-like and the time-like edges. We will not use the light-like edges.

The CM metric is now given by the same expression as in the Euclidean case (2.2), while the physical PL metric is given by

$$g_{\mu\nu}(\sigma) = \frac{G_{\mu\nu}(\sigma)}{2|L_{0\mu}||L_{0\nu}|}, \qquad (2.12)$$

where the modulus accounts for the fact that now $L_\epsilon$ can be a complex number.

In order to ensure the Minkowski signature of the PL metric we need to impose

$$\det G(\sigma) < 0, \qquad (2.13)$$

for any four-simplex $\sigma$ in $T(M)$. This is analogous to the first restriction in the Euclidean case (2.4). However, there is no need for the analogs of the second (2.5) and the third restriction (2.6), since the signatures of $\det G(\tau)$ and $\det G(\Delta)$ are not fixed in the Minkowski case. Namely, $\det G(\tau) > 0$ if $\tau$ belongs to a Euclidean hyper-plane of $g_{\mu\nu}(\sigma)$, while $\det G(\tau) < 0$ if $\tau$ belongs to a Minkowski hyper-plane. Also $\det G(\Delta) > 0$ if $\Delta$ belongs to a Euclidean plane while $\det G(\Delta) < 0$ if $\Delta$ belongs to a Minkowski plane.

The volumes of $n$-simplices can be defined as

$$(V_n)^2 = \frac{|\det G_n|}{2^n(n!)^2} > 0, \qquad n = 2, 3, 4, \qquad (2.14)$$

so that $V_n > 0$. Note that in the $n = 1$ case we should distinguish between the labels $L_\epsilon \in \mathbb{C}$ and their volumes $|L_\epsilon| > 0$. We will also use an equivalent labeling $L_\epsilon \to |L_\epsilon|$ with an indication $S$ or $T$ for an edge $\epsilon$.

Given that the edge lengths can take real or imaginary values in a Minkowski space, this implies that the angles between the vectors can be real or complex. Let us consider the angles in a Minkowski plane. Such angles can be defined as

$$\cos\alpha = \frac{\vec{u} \cdot \vec{v}}{||\vec{u}||\,||\vec{v}||}, \qquad \sin\alpha = \sqrt{1 - \cos^2\alpha}, \qquad \alpha \in \mathbb{C}, \qquad (2.15)$$

where $\vec{u} = (u_1, u_0)$, $\vec{u} \cdot \vec{v} = u_1 v_1 - u_0 v_0$ and $||\vec{u}|| = \sqrt{\vec{u} \cdot \vec{u}}$.

Consider two space-like vectors $\vec{u} = (1,0)$ and $\vec{v} = (\cosh a, \sinh a)$, $a \in \mathbb{R}$. Since $||\vec{u}|| = ||\vec{v}|| = 1$ then

$$\cos\alpha = \cosh a, \quad \sin\alpha = i\sinh a \quad \Rightarrow \quad \alpha = ia. \tag{2.16}$$

In the case of a space-like vector $\vec{u} = (1,0)$ and a time-like vector $\vec{v} = (\sinh a, \cosh a)$, we have $||\vec{u}|| = 1$ and $||\vec{v}|| = i$, so that

$$\cos\alpha = -i\sinh a, \quad \sin\alpha = \cosh a \quad \Rightarrow \quad \alpha = \frac{\pi}{2} - ia. \tag{2.17}$$

And if we have two time-like vectors $\vec{u} = (0,1)$ and $\vec{v} = (\sinh a, \cosh a)$, then

$$\cos\alpha = \cosh a, \quad \sin\alpha = i\sinh a \quad \Rightarrow \quad \alpha = ia. \tag{2.18}$$

The definition (2.15) then implies that the sum of the angles between two intersecting lines in a Minkowski plane is $2\pi$.

In order to define the dihedral angles in the Minkowski case we will introduce

$$(v_n)^2 = \frac{\det G_n}{2^n (n!)^2}, \quad n = 2, 3, 4, \tag{2.19}$$

so that $v_n = V_n$ for $\det G_n > 0$ or $v_n = iV_n$ for $\det G_n < 0$. In the $n = 1$ case we have $v_\epsilon = L_\epsilon$ for a space-like edge or $v_\epsilon = iL_\epsilon$ for a time-like edge where $L_\epsilon > 0$. Then the angle between two edges in a triangle is given by

$$\sin\alpha_\pi^{(\Delta)} = \frac{2\,v_\Delta}{v_\epsilon\,v_{\epsilon'}}, \tag{2.20}$$

where $\pi$ the common point (known as the hinge).

The dihedral angle between two triangles sharing an edge in a tetrahedron is given by

$$\sin\phi_\epsilon^{(\tau)} = \frac{3}{2}\frac{v_\epsilon\,v_\tau}{v_\Delta\,v_{\Delta'}}, \tag{2.21}$$

while the dihedral angle between two tetrahedrons sharing a triangle in a 4-simplex is given by

$$\sin\theta_\Delta^{(\sigma)} = \frac{4}{3}\frac{v_\Delta\,v_\sigma}{v_\tau\,v_{\tau'}}. \tag{2.22}$$

The formulas (2.20), (2.21) and (2.22) are generalizations the corresponding Euclidean formulas such that $V_n \to v_n$, and the novelty in the Minkowski case is that $\sin\theta$ is not restricted to the interval $[-1,1]$, but $\sin\theta \in \mathbb{R}$ or $\sin\theta \in i\mathbb{R}$. This also means that the Minkowski dihedral angles can take the complex values.

In the case of a dihedral angle $\theta_\Delta^{(\sigma)}$ there are two possibilities. If the triangle $\Delta$ is in a Minkowski (ST) plane, then $\theta$ will be an angle in an orthogonal Euclidean (SS) plane, so that $\sin\theta = \sin a$. If $\Delta$ is in an SS plane, then $\theta$ will be in an orthogonal ST plane, so that $\sin\theta = \cosh a$ or $\sin\theta = i\sinh a$.

The deficit angle will then take the following values:

$$\delta_\Delta = 2\pi - \sum_{\sigma \supset \Delta} \theta_\Delta^{(\sigma)} \in \mathbb{R}\,, \tag{2.23}$$

when $\Delta$ is an ST triangle, while

$$\delta_\Delta = 2\pi - \sum_{\sigma \supset \Delta} \theta_\Delta^{(\sigma)} \in \frac{\pi}{2}\mathbb{Z} + i\mathbb{R}\,, \tag{2.24}$$

when $\Delta$ is an SS triangle. Note that for an SS triangle the triangle inequalities are valid, while for an ST triangle they do not apply.

The appearance of the complex values for the deficit angles in the Minkowski signature case raises the question of how to generalize the Euclidean Regge action such that the new action is real. A proposal for a Lorentzian Regge action was given in [3]

$$S_R = \sum_{\Delta \in SS} A_\Delta \frac{1}{i} \delta_\Delta + \sum_{\Delta \in ST} A_\Delta \delta_\Delta\,. \tag{2.25}$$

However, the problem with this definition is that *a priori* $\tilde{S}_R \in \mathbb{R} + i\frac{\pi}{2}\mathbb{Z}$, so that one has to verify for a given triangulation that $Im\,S_R = 0$.

In order to avoid this difficulty, we will take

$$S_R = Re\left(\sum_{\Delta \in SS} A_\Delta \frac{1}{i} \delta_\Delta\right) + \sum_{\Delta \in ST} A_\Delta \delta_\Delta\,. \tag{2.26}$$

This definition can be justified by the fact that the authors of [3] have verified that $Im\,S_R = 0$ for a special class of triangulations, which are physically relevant, and they are called the casual triangulations.

Note that one can avoid the problem of the complex Regge action by using the triangulations which contain only the ST triangles or triangulations which contain only the SS triangles. In the ST case the Regge action is given by the second term in (2.26), which is real. In the SS case, the Regge action is given by the first term, which is real. This was the approach taken in [88], since any cylindrical 4-manifold can be triangulated in this way.

## 2.2    Coupling of matter in Regge GR

Given a scalar field $\phi(x)$ on $M$, we introduce on $T(M)$ a discrete set of values $\{\phi(\pi) \,|\, \pi \in T(M)\}$, where $\pi$ is a vertex of $T(M)$. Let $\pi_0, \pi_1, \ldots, \pi_4$ be the vertices of a 4-simplex $\sigma$ and let $\mu = 1, 2, 3, 4$, then

$$\partial_\mu \phi(x) \to \Delta_\mu \phi(\sigma) = \frac{\phi(\pi_\mu) - \phi(\pi_0)}{L_{0\mu}}\,,$$

$$\int_M d^4x \sqrt{-g}\, g^{\mu\nu}\, \partial_\mu \phi \partial_\nu \phi \to \sum_{\sigma \in T(M)} V_\sigma(L) g_\sigma^{\mu\nu}(L) \Delta_\mu \phi \Delta_\nu \phi\,,$$

$$\int_M d^4x \sqrt{-g}\, U(\phi) \to \sum_{\pi \in T(M)} U(\phi_\pi) V_\pi^*(L)\,,$$

where $V_\pi^*(L)$ is the volume of the dual 4-cell around a vertex $\pi$ and $U(\phi)$ is the scalar field potential function. Also $L_{0\mu} = |L_\epsilon|$, i.e. they are positive real numbers, and we obtain

$$S_{sclr}(\phi, L) = \frac{1}{2} \sum_{\sigma \in T(M)} V_\sigma(L) g_\sigma^{\mu\nu}(L) \Delta_\mu \phi \Delta_\nu \phi - \sum_{\pi \in T(M)} U(\phi_\pi) V_\pi^*(L)\,.$$

$$(2.27)$$

For a gauge field $A_\mu(x) = A_\mu^I(x) T_I$, where $T_I$ is a basis of a Lie algebra, we have

$$\partial_\mu A_\nu(x) \to \Delta_\mu A_\nu(\sigma) = \frac{A_\nu(\pi_\mu) - A_\nu(\pi_0)}{L_{0\mu}}\,,$$

$$F_{\mu\nu}(x) \to F_{\mu\nu}(\sigma) = \Delta_{[\mu} A_{\nu]}(\sigma) + [A_\mu(\pi_0), A_\nu(\pi_0)]\,,$$

so that the Yang-Mills action

$$S_{YM} = \int_M d^4x \sqrt{-g}\, g^{\mu\rho} g^{\nu\lambda}\, Tr\left(F_{\mu\nu} F_{\rho\lambda}\right)$$

becomes

$$S_{YM}(A, L) = \sum_{\sigma \in T(M)} V_\sigma(L)\, g_\sigma^{\mu\rho} g_\sigma^{\nu\lambda}\, Tr\left(F_{\mu\nu}(\sigma) F_{\rho\lambda}(\sigma)\right)\,. \qquad (2.28)$$

Given a Dirac spinor $\psi^\alpha(x)$ on $M$, we are interested in the Dirac action, see section 2.5.2. In order to discretize the Dirac action on $T(M)$, aside from values $\psi(\pi)$, we will also need the values $\psi(v)$, where $v$ are the vertices of the dual triangulation $T^*(M)$. Here we will take that the components $\psi^\alpha$ are complex numbers, since we are dealing with the classical theory.[1]

In order to construct the Dirac action on $T(M)$, one has to introduce the tetrads as 4-vectors $e^a{}_\mu(\sigma)$ in $\sigma$ such that

$$e^a{}_\mu(\sigma)e_{a\nu}(\sigma) = g^{(\sigma)}_{\mu\nu}(L).$$

Then the spin connection on $T(M)$ can be defined by using $SO(1,3)$ matrices $\Lambda(l)$ associated to dual edges $l \in T^*(M)$, such that

$$e^a{}_\mu(\sigma') = \Lambda^a{}_b(\sigma,\sigma')e^b{}_\mu(\sigma).$$

Here the 4-simplices $\sigma$ and $\sigma'$ share a tetrahedron $\tau$, which is dual to the edge $l$. The matrix $\Lambda$ can be understood as the holonomy of a connection along the dual edge $l$.

One can then calculate the holonomy for a closed loop $\partial f$ of a dual face $f$ associated to a triangle $\Delta$. It is given by

$$Hol(f) = \prod_{l \in \partial f} \Lambda(l),$$

and when the area $A_f$ of the face $f$ tends to zero, we have

$$Tr\left(Hol(f) - Id\right) \approx \delta_f A_f,$$

where $\delta_f = \delta_\Delta$ is the deficit angle. We can then associate a scalar curvature $R_f = \delta_f/A_f$ to the face $f$, or equivalently, $R_f$ can be associated to the dual triangle $\Delta$.

From these results one can easily derive the Regge action, since

$$\int_M \sqrt{-g}\,d^4x\,R(g) \to \sum_{\Delta \in T(M)} A_\Delta A_f \frac{\delta_\Delta}{A_f} = \sum_{\Delta \in T(M)} A_\Delta \delta_\Delta,$$

where $A_\Delta A_f$ is the 4-volume of the diamond-like region spanned by a triangle $\Delta$ and the dual face $f$, while the corresponding scalar curvature is $\delta_\Delta/A_f$.

A spin connection $\omega(l)$ can be associated to a matrix $\Lambda(l)$ by using

$$\Lambda(l) = \exp\int_l \omega_\mu dx^\mu \Rightarrow \Lambda^b{}_a(l) \approx \delta^b_a + L_l\,\omega^b{}_a(l),\ \text{for}\ L_l \to 0,$$

where $L_l$ is the length of a dual edge $l$.

---

[1]One could in principle retain those complex-number values even when defining the path integral for spinors. However, in this approach there will appear the determinant factors in the path integral measure which come from the second-class constraints of the Dirac spinor action. It is easier to keep track of those determinants by the trick of considering the spinor components as the elements of a Grassmann algebra and by defining the operation of integration on a Grassmann algebra.

In the case of the kinetic term of the Dirac action we have

$$\int_M \varepsilon^{abcd} e_a \wedge e_b \wedge e_c \wedge \bar\psi \gamma_d \, \mathrm{d}\psi = \int_M d^4x \, \varepsilon^{abcd} e_{a\mu} e_{b\nu} e_{c\rho} \, \bar\psi \gamma_d \, (\partial_\lambda \psi) \, \varepsilon^{\mu\nu\rho\lambda}$$

so that on $T(M)$ we obtain

$$S_D^{(1)}(\psi, L) = \sum_{\sigma \in T(M)} V_\sigma(L) \, \varepsilon^{abcd} e_{a\mu}(\sigma) e_{b\nu}(\sigma) e_{c\rho}(\sigma) \, \bar\psi(\pi_0) \gamma_d \, \Delta_\lambda \psi(\sigma) \, \varepsilon^{\mu\nu\rho\lambda} \,,$$

$$(2.29)$$

where

$$\Delta_\lambda \psi^\alpha(\sigma) = \frac{\psi^\alpha(\pi_\lambda) - \psi^\alpha(\pi_0)}{L_{0\lambda}} \,,$$

and $\pi_0$ and $\pi_\lambda$ are the vertices of $\sigma$. For the second term of the Dirac action

$$S_D^{(2)} = \int_M \varepsilon^{abcd} e_a \wedge e_b \wedge e_c \wedge \bar\psi \{\gamma_d, \omega\} \psi \,,$$

we obtain on $T(M)$

$$S_D^{(2)}(\psi, L) = \sum_{\tau \in T(M)} \varepsilon^{abcd} B_{abc}(\tau) \, L_l \, \bar\psi(v) \{\gamma_d, \omega(l)\} \psi(v') \,, \qquad (2.30)$$

where

$$B_{abc}(\tau) = V_\tau(L) e_{ai}(\tau) e_{bj}(\tau) e_{ck}(\tau) \varepsilon^{ijk} \,,$$

$\omega(l) = \omega^{ab}(l)\gamma_{ab}$, while $v$ and $v'$ are the vertices of the dual edge $l$. The mass term in the Dirac action becomes

$$S_D^{(3)}(\psi, L) = \sum_{\sigma \in T(M)} V_\sigma(L) \bar\psi(v) \psi(v) \,, \qquad (2.31)$$

or we can also use

$$\tilde{S}_D^{(3)}(\psi, L) = \sum_{\pi \in T(M)} V_\pi^*(L) \bar\psi(\pi) \psi(\pi) \,. \qquad (2.32)$$

Note that for spinors in a curved spacetime we need two sets of values in order to define the Dirac action on a PL manifold: a set of values on the vertices of $T(M)$ and a set of values on the vertices of $T^*(M)$. In flat space $\omega = 0$, and we need just one set of the values. In the case of scalars and vectors, even in the curved spacetime, only one set of values is sufficient to define the action.

## 2.3   Area-Regge action

The fundamental variables in the Regge formulation of GR are the edge lengths $L_\epsilon$ in a triangulation $T(M)$. Due to the development of the SF models, where the fundamental variables are the spins of the triangles and the corresponding intertwiners for the tetrahedrons, one is interested in formulations of GR where the areas of triangles $A_\Delta$ are the fundamental variables, since the areas and the spins in an SF model are related by

$$A_\Delta = \left( \sqrt{j_\Delta(j_\Delta + c)} + c' \right) l_P^2 \,,$$

where $c$ and $c'$ are normal ordering constants. The usual choices are $(c, c') = (1, 0)$, $(1/2, 0)$ or $(0, 1/2)$.

Given the 10 areas in a 4-simplex $\sigma$, we can determine the 10 edge lengths, provided that there is a solution of the system of 10 equations

$$A_{ijk} = \sqrt{s_{ijk}(L_{ij} - s_{ijk})(L_{jk} - s_{ijk})(L_{ki} - s_{ijk})} \,, \quad i, j, k \in \sigma \,, \qquad (2.33)$$

where

$$s_{ijk} = \frac{L_{ij} + L_{jk} + L_{ki}}{2} \,.$$

Let us assume that we can do this for all 4-simplices in a triangulation, i.e.

$$L_\epsilon^{(\sigma)} = f_\epsilon(A^{(\sigma)}) \,, \quad \sigma \in T(M) \,. \qquad (2.34)$$

Then we can rewrite the Regge action as

$$S_R(L) = \sum_{\Delta \in T(M)} A_\Delta(L)\, \delta_\Delta(L) = \sum_{\Delta \in T(M)} A_\Delta\, \tilde{\delta}_\Delta(A) \equiv \tilde{S}_R(A) \,, \qquad (2.35)$$

where

$$\tilde{\delta}_\Delta(A) = \delta_\Delta(f(A)) \,.$$

The new action is called the area-Regge action, and one can consider a theory of gravity based on it. One can show that a variation of $\tilde{S}_R(A)$ with respect to $A_\Delta$ gives the equation of motion (EoM)

$$\tilde{\delta}_\Delta(A) = 0 \,, \qquad (2.36)$$

see [16]. The equation (2.36) seem to imply that all deficit angles vanish, and one may conclude that the only solutions are the zero-curvature solutions, i.e. flat spacetimes. However, this is not a correct conclusion, because an arbitrary set of triangle areas does not define a metric geometry in $T(M)$. This means that it is not possible to find a set of the edge lengths which satisfy the triangle inequalities such that the every area is given by the Heron formula (2.33).

In order to have a metric geometry, the solutions (2.34) of the systems of equations (2.33) have to match at the border of two 4-simplices. Hence we need to impose the constraints

$$f_\epsilon(A^{(\sigma)}) = f_\epsilon(A^{(\sigma')}), \quad \epsilon \in \sigma \cap \sigma', \tag{2.37}$$

in order to have a theory for metric geometries [75]. For example, in the case of a triangulation of a 4-sphere consisting of 6 vertices, 15 edges and 20 triangles, we need five independent relations of the constraints (2.37) to be imposed on the 20 triangle areas in order to have a metric geometry.

Therefore without the edge-length constraints (2.37) we do not have a metric geometry, so that the meaning of the equation (2.36) is not the same as the vanishing of the curvature for each triangle.

### 2.3.1 *Area-angle action*

Note that in order to define the area-Regge action we had to assume that the system of equations (2.33) had a solution for each 4-simplex. If one takes an arbitrary assignment of the areas, the system (2.33) will not have a solution, so that one has to introduce new constraints. One can do this by imposing the constraints (2.37), but a more elegant way is to introduce the variables for the dihedral angles in a tetrahedron and the corresponding constraints, see [42].

The motivation for introducing the tetrahedron dihedral angles as dynamical variables comes from the spin foam models of quantum gravity, since the tetrahedron dihedral angles are equivalent to the unit normal vectors for the triangles of a tetrahedron, and these are related to the intertwiners in a spin foam, see section 3.2. Hence another reason for using the tetrahedron dihedral angles is to understand the classical dynamics of spin foams.

Given a 4-simplex $\sigma$, with areas $A_\Delta$ and the dihedral angles $\phi_\epsilon^{(\tau)}$, the constraints which those variables have to satisfy in order that $\sigma$ has a metric geometry, are

$$A_\Delta = \sum_{\Delta' \in \tau} A_{\Delta'} \cos \phi_{\Delta \cap \Delta'}^{(\tau)}, \quad \Delta \in \tau, \tag{2.38}$$

and

$$\alpha_\nu^{(\Delta)}(\phi) = \alpha_\nu^{(\Delta)}(\tilde\phi), \tag{2.39}$$

where $\alpha(\phi)$ is the dihedral angle for a vertex $\pi$ in a triangle $\Delta$ expressed as a function of the dihedral angles $\phi$ in a tetrahedron $\tau$, while $\alpha(\tilde\phi)$ is the same angle expressed in terms of the dihedral angles of a tetrahedron $\tilde\tau$ that shares the triangle $\Delta$ with $\tau$.

There are 10 independent constraints (2.38). This can be seen from the alternative form of these constraints, which is

$$\sum_{\Delta \in \tau} A_\Delta \vec{n}_{\Delta,\tau} = 0 \,, \tag{2.40}$$

where $\vec{n}$ are the unit vectors normal to the tetrahedron triangles, and

$$\cos\phi_\epsilon^{(\tau)} = -\vec{n}_{\Delta,\tau} \cdot \vec{n}_{\Delta',\tau} \,, \quad \Delta \cap \Delta' = \epsilon \,.$$

Since each $\vec{n}$ has two independent components, from (2.40) we see that there are $2 \cdot 5 = 10$ constraints in a 4-simplex. There are $2 \cdot 10 = 20$ independent constraints (2.39), because of the 3 constraints (2.39) per triangle, only two are independent, since $\alpha + \alpha' + \alpha'' = \pi$. Since we have $5 \cdot 6 = 30$ dihedral angles $\phi$, the number of independent variables is

$$n^*(\sigma) = 10 + 30 - 10 - 20 = 10 \,,$$

which correspond to the 10 edge lengths in a 4-simplex.

Therefore

$$S(A,\phi) = \sum_\Delta A_\Delta \delta_\Delta(\phi) - \sum_\tau \sum_{\Delta \in \tau} \lambda_{\Delta,\tau} C_{\Delta,\tau} - \sum_\sigma \sum_{\Delta \in \sigma} \sum_{\nu \in \Delta} \mu_{\nu,\Delta}^{(\sigma)} C_{\nu,\Delta}^{(\sigma)} \,,$$

where $\lambda$ and $\mu$ are the Lagrange multipliers enforcing the constraints

$$C_{\Delta,\tau} = A_\Delta - \sum_{\Delta' \in \tau} A_{\Delta'} \cos\phi_{\Delta \cap \Delta'}^{(\tau)} \,, \tag{2.41}$$

and

$$C_{\nu,\Delta}^{(\sigma)} = \alpha_\nu^{(\Delta)}(\phi^{(\tau)}) - \alpha_\nu^{(\Delta)}(\phi^{(\tau')}) \,, \quad \Delta = \tau \cap \tau', \sigma = \tau \cup \tau' \,. \tag{2.42}$$

For example, in the triangulation of a 4-sphere consisting of 6 vertices, 15 edges and 20 triangles, there are 15 tetrahedrons. Therefore we have $6 \cdot 15 = 90$ dihedral angles $\phi$. There are $2 \cdot 15 = 30$ tetrahedron closure constraints and $2 \cdot 10 \cdot 6 = 120$ triangle dihedral angle constraints. We have 110 variables and 150 constraints, which means that some of the constraints are not independent. Out of 150 area-angle constraints (2.41) and (2.42), there are 90 independent constraints, which can be seen in the following way. Let us impose 30 area-angle constraints on the 4-simplex $\sigma_{12345}$. This will determine the 10 edge lengths $L_{ij}$ in this 4-simplex. By imposing another 30 area-angle constraints for $\sigma_{12346}$, we will determine 4 new edge lengths $L_{k6}$, and by imposing another 30 area-angle constraints on $\sigma_{23456}$ we will determine the last edge length, $L_{56}$. The rest of the area-angle constraints will not determine any new edge lengths.

Since each pair of these 4-simplices has 6 common edges, we can impose 5 edge-length constraints (2.37), so that the number of independent DoF is

$$n^* = 110 - 90 - 5 = 15 \,,$$

which corresponds to the 15 edge lengths. Since the 20 areas in the triangulation are real numbers, the 15 edge lengths satisfy the triangle inequalities and hence we can define a metric on the corresponding PL manifold.

## 2.4   $BF$ theory

One of the simplest and most commonly studied topological field theories is the so-called $BF$ theory [6, 21, 23, 24, 28, 29, 38, 39, 63, 116]. It represents the starting point for the construction of various other field theories, and its topological nature allows one to explicitly construct a corresponding quantum theory in a fully non-perturbative manner, by using the path-integral techniques, see chapter 3. In addition, the techniques from category theory, specifically higher gauge theory (HGT) [8], enable one to generalize the $BF$ theory to its higher-category analogues, which are called $nBF$ theories, see section 2.5.

Given a Lie group $G$ and a $D$-dimensional manifold $M_D$, one defines the action of a $BF$ theory as

$$S_{BF}[A, B] = \int_{M_D} \langle B \wedge F \rangle_{\mathfrak{g}} \,. \tag{2.43}$$

The name "$BF$ theory" stems[2] from the forms $B$ and $F$ which enter the action (2.43). Here $F = \mathrm{d}A + A \wedge A$ is the curvature 2-form of the 1-form $A$ representing the connection over a principal bundle $G \to M_D$, while $B$ is the Lagrange multiplier $(D - 2)$-form. Both $A$ and $B$ are $\mathfrak{g}$-valued differential forms, where $\mathfrak{g}$ is a Lie algebra of the Lie group $G$, and $\langle \_, \_ \rangle_{\mathfrak{g}}$ is a $G$-invariant symmetric non-degenerate bilinear form on $\mathfrak{g}$. The notation $\langle \_ \wedge \_ \rangle_{\mathfrak{g}}$ means that the two arguments, being differential forms, should be additionally multiplied with a wedge product.

---

[2]It is sometimes claimed that $BF$ is a shorthand for "background field", but the $BF$ action has nothing whatsoever to do with any background field (the Lagrange multiplier $B$ is *not* a background field since it is being varied over and has a corresponding EoM). The $BF$ theory also has little, if anything, to do with the background field method used in QFT.

For instance, if $B$ and $F$ are both $\mathfrak{g}$-valued 2-forms over a 4-dimensional manifold, $B, F \in \Lambda^2(M_4) \otimes \mathfrak{g}$, we have

$$B = \frac{1}{2}B^a{}_{\mu\nu}(x)\,\mathrm{d}x^\mu \wedge \mathrm{d}x^\nu \otimes T_a\,, \qquad F = \frac{1}{2}F^b{}_{\rho\sigma}(x)\,\mathrm{d}x^\rho \wedge \mathrm{d}x^\sigma \otimes T_b\,,$$

where $T_a \in \mathfrak{g}$ are the generators of the Lie algebra $\mathfrak{g}$, so that

$$\begin{aligned}
\langle B \wedge F \rangle_{\mathfrak{g}} &= \frac{1}{4}B^a{}_{\mu\nu}(x)F^b{}_{\rho\sigma}(x)\,\mathrm{d}x^\mu \wedge \mathrm{d}x^\nu \wedge \mathrm{d}x^\rho \wedge \mathrm{d}x^\sigma\,\langle T_a, T_b \rangle_{\mathfrak{g}} \\
&= \frac{1}{4}B^a{}_{\mu\nu}(x)F^b{}_{\rho\sigma}(x)\,\varepsilon^{\mu\nu\rho\sigma}d^4x\,g_{ab}\,,
\end{aligned}$$

where $g_{ab} \equiv \langle T_a, T_b \rangle_{\mathfrak{g}}$ are the components of the bilinear form in the given basis of the Lie algebra.

The action (2.43) can then be rewritten as

$$S_{BF}[A, B] = \int_{M_D} d^4x\,\frac{1}{4}B^a{}_{\mu\nu}(x)F^b{}_{\rho\sigma}(x)\,\varepsilon^{\mu\nu\rho\sigma}g_{ab}\,, \qquad (2.44)$$

form which one can read off the Lagrangian density of the $BF$ theory as

$$\mathcal{L}_{BF} = \frac{1}{4}B^a{}_{\mu\nu}(x)F^b{}_{\rho\sigma}(x)\,\varepsilon^{\mu\nu\rho\sigma}g_{ab}\,.$$

By expanding the $\mathfrak{g}$-valued 1-form $A$ into the basis as

$$A = A^a{}_\mu(x)\,\mathrm{d}x^\mu \otimes T_a\,,$$

and by using the relation $F = \mathrm{d}A + A \wedge A$, one can evaluate the curvature components $F^a_{\mu\nu}$ to be

$$F^a{}_{\mu\nu} = \partial_\mu A^a{}_\nu - \partial_\nu A^a{}_\mu + f^a{}_{bc}A^b{}_\mu A^c{}_\nu\,,$$

where $f^a{}_{bc}$ are the structure constants of the Lie algebra $\mathfrak{g}$ in the basis $T_a$,

$$[T_b, T_c] \equiv T_b T_c - T_c T_b = f^a{}_{bc}T_a\,.$$

In the case when $G$ is non-Abelian, the natural choice of the bilinear form is the Killing form,

$$g_{ab} = \mathrm{const}\cdot f^c{}_{ad}f^d{}_{bc}\,,$$

while in the Abelian case one can choose $g_{ab}$ freely, as long as one ensures that this choice is $G$-invariant, non-degenerate and symmetric.

One can easily show that the classical equations of motion of the theory are[3]

$$F = 0\,, \qquad \nabla B \equiv \mathrm{d}B + A \wedge^{[\,\cdot\,]} B = 0\,,$$

---

[3]The notation $\wedge^{[\,\cdot\,]}$ means that the two quantities are multiplied with the wedge product since they are differential forms, while simultaneously they are multiplied via the Lie bracket since they are elements of a Lie algebra. Later on we shall similarly introduce the more general product $\wedge^\triangleright$, where $\triangleright$ will be an arbitrary binary operation between given objects defined in a given context.

which means that the connection $A$ is a flat connection, while $B$ is constant (in a suitable gauge). Therefore, the theory appears locally trivial, having no local propagating degrees of freedom [31, 89–91]. However, the theory has global degrees of freedom, which come from the topological properties of the principal bundle $G \to M_D$.

As far as symmetries are concerned, the theory is manifestly invariant with respect to diffeomorphisms, since the action is expressed in terms of differential forms, which are manifestly diffeomorphism invariant. For example,

$$
\begin{aligned}
B \to B' &= \frac{1}{2} B'_{\mu\nu} \, \mathrm{d}x'^{\mu} \wedge \mathrm{d}x'^{\nu} = \frac{1}{2} \left( B_{\rho\sigma} \frac{\partial x^{\rho}}{\partial x'^{\mu}} \frac{\partial x^{\sigma}}{\partial x'^{\nu}} \right) \frac{\partial x'^{\mu}}{\partial x^{\lambda}} \mathrm{d}x^{\lambda} \wedge \frac{\partial x'^{\nu}}{\partial x^{\tau}} \mathrm{d}x^{\tau} \\
&= \frac{1}{2} B_{\lambda\tau} \, \mathrm{d}x^{\lambda} \wedge \mathrm{d}x^{\tau} = B \,,
\end{aligned}
$$

and similarly for any other differential form, or a tensor in general.

Second, the theory is invariant with respect to gauge symmetry $G$. For every $g \in G$, define the gauge transformation as

$$
A \to A' = gAg^{-1} + g\mathrm{d}g^{-1} \,, \qquad B \to B' = gBg^{-1} \,.
$$

From the expression $F = \mathrm{d}A + A \wedge A$ it follows that

$$
F \to F' = gFg^{-1} \,,
$$

so one obtains that the $BF$ action (2.43) remains invariant, owing to the $G$-invariance of the bilinear form,

$$
\langle gXg^{-1}, gYg^{-1} \rangle_{\mathfrak{g}} = \langle X, Y \rangle_{\mathfrak{g}} \,, \qquad \forall X, Y \in \mathfrak{g} \,.
$$

Finally, the theory is invariant with respect to the shift symmetry, defined as

$$
A \to A' = A \,, \qquad B \to B' = B + \nabla\lambda \,,
$$

where $\lambda$ is an arbitrary algebra-valued 1-form, $\lambda \in \Lambda^1(M_D) \otimes \mathfrak{g}$. Since $F' = F$, substituting the transformation into the action gives

$$
S_{BF} \to S'_{BF} = S_{BF} + \int_{M_D} \langle \nabla\lambda \wedge F \rangle_{\mathfrak{g}} \,.
$$

After a partial integration, the integral on the right-hand side vanishes (up to a boundary term), owing to the Bianchi identity $\nabla F \equiv 0$, leaving the action invariant.

As a final remark regarding the basic properties of the $BF$ theory, note that it does not require a metric on the manifold in order to be well defined. This property is very useful when dealing with reformulations and alternatives to general relativity, as we shall see below.

Despite its intrinsic simplicity, one can employ $BF$ theory (and its generalizations, see section 2.5) to construct other theories which are quite interesting and rich, as we shall see below. In addition, being topological, $BF$ theory can be quantized non-perturbatively by using the methods of topological quantum field theory (TQFT). This gives a rigorous definition of the $BF$ path integral,

$$Z_{BF} = \int \mathcal{D}A \int \mathcal{D}B \, e^{iS_{BF}[A,B]} \,,$$

which we will discuss in detail in chapter 3.

In order to define $Z_{BF}$ we will need a form of the action on a PL manifold $T(M)$. It is given by

$$S_{BF}^{T(M)} = \sum_{\Delta \in T(M)} Tr\left(B_\Delta F_f\right), \tag{2.45}$$

where

$$B_\Delta = \int_\Delta B \,, \quad F_f = \int_f F \,,$$

and $f$ is a face of the dual triangulation $T^*(M)$ which corresponds to a triangle $\Delta$ of $T(M)$, i.e. the vertices of $f$ are inside the 4-simplices which share the triangle $\Delta$.

### 2.4.1  *3D Palatini action and Chern-Simons theory*

The most interesting examples of a $D = 3$ $BF$ theory are the cases when $G = SO(3)$ or $G = SO(2,1)$. They describe three-dimensional GR for Euclidean and Minkowski metrics, respectively.

If we denote a basis of the corresponding Lie algebra $\mathfrak{g}$ as $L_a$, with the group index taking values $a - 1, 2, 3$, we can write the commutation relations, structure constants and the Killing form as

$$[L_b, L_c] = f^a{}_{bc}(\zeta)L_a \,, \qquad \eta_{ab} = \begin{bmatrix} \zeta & 0 & 0 \\ 0 & 1 & 0 \\ 0 & 0 & 1 \end{bmatrix} \,, \qquad \zeta = \pm 1 \,.$$

The parameter $\zeta$ is equal to $+1$ or $-1$, for the cases of $SO(3)$ or $SO(2,1)$, respectively. The bilinear form $\eta_{ab}$ and its inverse $\eta^{ab}$ are used to raise and lower group indices.

The $BF$ action can be then written as

$$S_{BF}[e, \omega] = \int_{M_3} e_a \wedge F^a(\omega) \,,$$

where we have conspicuously relabeled the 1-form $B$ into $e$ (suggesting the interpretation of a triad), and the connection 1-form $A$ into $\omega$ (suggesting the interpretation of the spin connection).

This action has a form of a 3-dimensional Palatini action for general relativity, and we will show that it is equivalent to the 3-dimensional Einstein-Hilbert action. To see this, expand the forms $e^a$ and $F^a$ into a basis, and substitute them into the action to recover the traditional tensor notation:

$$S_{BF} = \int_{M_3} e_{a\mu} F^a{}_{\rho\sigma} \, \mathrm{d}x^\mu \wedge \mathrm{d}x^\rho \wedge \mathrm{d}x^\sigma = \int_{M_3} e_{a\mu} F^a{}_{\rho\sigma} \, \varepsilon^{\mu\rho\sigma} d^3x \, .$$

Then apply the identity

$$F^a{}_{\rho\sigma} = \varepsilon^{abc} e_{b\lambda} e_{c\nu} R^{\lambda\nu}{}_{\rho\sigma}$$

to rewrite the curvature tensor in its traditional form, and reorganize the terms in the action as

$$S_{BF} = \int_{M_3} \left[ \varepsilon_{abc} e^a{}_\mu e^b{}_\lambda e^c{}_\nu \right] \varepsilon^{\mu\rho\sigma} R^{\lambda\nu}{}_{\rho\sigma} \, d^3x \, .$$

Next recognize that the term in the brackets is related to the determinant $e \equiv \det(e^a{}_\mu)$ of the triad via the identity

$$\varepsilon_{abc} e^a{}_\mu e^b{}_\lambda e^c{}_\nu = e \, \varepsilon_{\mu\lambda\nu} \, ,$$

and that the contraction of the two totally antisymmetric Levi-Civita symbols is given as

$$\varepsilon_{\mu\lambda\nu} \varepsilon^{\mu\rho\sigma} = \zeta \left( \delta^\rho_\lambda \delta^\sigma_\nu - \delta^\sigma_\lambda \delta^\rho_\nu \right) \, .$$

Note that this contraction depends on the choice of the signature $\zeta$. Substituting these into the action, and recalling that the curvature scalar is defined as a double contraction of the Riemann tensor, $R = R^{\rho\sigma}{}_{\rho\sigma}$, after a little algebra we obtain

$$S_{BF} = 2\zeta \int_{M_3} e R \, d^3x \, .$$

As a final step, employ the triads to introduce the spacetime metric in the standard way,

$$g_{\mu\nu} = e^a{}_\mu e^b{}_\nu \eta_{ab} \, .$$

Taking the determinant of both sides, denoting $g \equiv \det(g_{\mu\nu})$, and noting that $\det(\eta_{ab}) = \zeta$, one obtains that $g = \zeta e^2$, from which it follows that $|e| = \sqrt{\zeta g}$. Under the standard assumption that $e > 0$ everywhere on the manifold, the $BF$ action finally obtains the form

$$S_{BF} = 2\zeta \int_{M_3} d^3x \sqrt{\zeta g} \, R = S_{\mathrm{EH}} \, .$$

In other words, we have demonstrated that the action for general relativity in $3D$ is precisely the $BF$ action. This is a remarkable property of the $BF$ theory, which unfortunately holds only in three dimensions. The underlying reason for this is that $D = 3$ gravity has no propagating local degrees of freedom, while in $D \geqslant 4$ gravity has a non-zero number of those, and $BF$ theory has none. Therefore, $BF$ theory can in principle precisely describe gravity only in $D = 3$, and this is indeed the case as demonstrated above. While suggesting that $BF$ theory can be useful as a starting point also in higher-dimensional cases, these will require a deformation of the $BF$ action in order to reproduce Einstein field equations. This topic will be tackled in subsection 2.4.2 and further in section 2.5.

The three-dimensional $SO(3)/SO(2,1)$ $BF$ theory has one more important property. We have seen that it is equal to the Palatini action, albeit without a potential cosmological constant. However, adding the cosmological constant can be achieved by a simple deformation of the $BF$ action, obtained by introducing the a second term proportional to the cosmological constant $\Lambda$, so that

$$S_{BF\Lambda}[e, \omega] = \int_{M_3} e_a \wedge F^a(\omega) - \frac{\Lambda}{3}\varepsilon_{abc} e^a \wedge e^b \wedge e^c. \qquad (2.46)$$

It is easy to see that this action can be rewritten as

$$S_{BF\Lambda} = 2\zeta \int_{M_3} d^3x \sqrt{\zeta g} \, (R - \Lambda),$$

and that the equations of motion give a spacetime of constant curvature,

$$R_{\mu\nu\rho\sigma} = \Lambda \left( g_{\mu\rho} g_{\nu\sigma} - g_{\mu\sigma} g_{\nu\rho} \right),$$

which satisfies the Einstein's field equation with a cosmological constant term

$$R_{\mu\nu} - \frac{1}{2} R \, g_{\mu\nu} + \Lambda \, g_{\mu\nu} = 0.$$

There is a very interesting relationship between the action (2.46) and the Chern-Simons theory [77], defined by the action

$$S_{CS}[A] = \int_{M_3} A_a \wedge \mathrm{d}A^a + \frac{2}{3}\varepsilon_{abc} A^a \wedge A^b \wedge A^c,$$

where $A$ is the connection 1-form for a Lie group $G$, in our case $SO(3)$ or $SO(2,1)$. It is also a topological theory, defined on 3-dimensional manifolds, with no local propagating degrees of freedom, while the global properties are determined by the topology of the principal bundle $G \to M_3$, as in the $BF$

theory case. However, it turns out that the action for the deformed $BF\Lambda$ theory (2.46) can be rewritten as a difference between two Chern-Simons actions, up to a boundary term.

In order to see this, let us introduce two different copies of the Chern-Simons theory, one for the connection 1-form $A^a$ and the other for the connection 1-form $\bar{A}^a$, both for the same principal bundle on which (2.46) is defined. Note that (2.46) consists of two 1-form fields — the Lagrange multiplier $e^a$ and the spin connection $\omega^a$. Employ this to introduce a change of variables of the form

$$\omega^a = A^a + \bar{A}^a, \qquad e^a = \frac{1}{\sqrt{-2\Lambda}}\left(A^a - \bar{A}^a\right),$$

with the inverse transformation being

$$A^a = \frac{1}{2}\left(\omega^a + \sqrt{-2\Lambda}e^a\right), \qquad \bar{A}^a = \frac{1}{2}\left(\omega^a - \sqrt{-2\Lambda}e^a\right).$$

We assume that the cosmological constant $\Lambda$ is negative, in order to have real connections $A$ and $\bar{A}$. Then by substituting the expressions for $\omega$ and $e$ into the action (2.46), after some algebra, one finds that

$$S_{BF\Lambda}[e,\omega] = \frac{1}{\sqrt{-2\Lambda}}\left(S_{CS}[A] - S_{CS}[\bar{A}] - \int_{\partial M_3} A^a \wedge \bar{A}_a\right). \qquad (2.47)$$

This result is fascinating, since it enables one to resolve the problem of quantization of general relativity in $D = 3$ by performing the quantization of the Chern-Simons theory, which is far simpler and well-studied. Of course, the relation (2.47) depends on the choices $D = 3$ and $\Lambda < 0$, so it is of limited value for a realistic theory of gravity. Nevertheless, one could study it as a toy model of a more realistic case, and hope that some of its properties may continue to hold in $D = 4$.

As far as the form of the $BF$ action on $T(M_3)$, we have

$$\int_M Tr\,(B \wedge F) \rightarrow \sum_{\epsilon \in T(M)} Tr\,(B_\epsilon F_f),$$

where

$$B_\epsilon = \int_\epsilon B, \qquad F_f = \int_f F$$

and $f$ is a dual triangulation face which corresponds to an edge $\epsilon$, i.e. the vertices of $f$ are inside the tetrahedrons which share the edge $\epsilon$.

On the other hand, for the cubic term from (2.46), we have

$$\int_M \varepsilon_{abc}\,e^a \wedge e^b \wedge e^b \rightarrow \sum_{\tau \in T(M)} \sum_{\epsilon,\epsilon',\epsilon'' \in \tau} \varepsilon_{abc}\,e^a_\epsilon e^b_{\epsilon'} e^c_{\epsilon''}.$$

### 2.4.2 *4D Plebanski action and general relativity*

In four spacetime dimensions, general relativity has two local propagating degrees of freedom, so that it is not a topological theory. As a consequence, it cannot be described by a $BF$ action, in contrast to the 3-dimensional case. Nevertheless, it is still possible to introduce a deformation of the $BF$ theory so that it does become equivalent to general relativity, a result due to Plebanski [106].

The Plebanski action can be constructed as follows. Start from a $BF$ theory in $D = 4$, with the gauge group $G = SO(4)$ or $G = SO(3,1)$, corresponding to the Euclidean or Minkowski cases, respectively. Denote the generators of the corresponding Lie algebra as $J_{ab}$, where the group index is the antisymmetrized pair of single indices $a, b \in \{0, 1, 2, 3\}$. The commutation relations and the Killing form are given as

$$[J_{ab}, J_{cd}] = f^{ef}{}_{ab,cd} J_{ef} \,, \qquad g_{ab,cd} = \frac{1}{2} \left( \eta_{ac}\eta_{bd} - \eta_{ad}\eta_{bc} \right) \,,$$

where the structure constants are

$$f^{ef}{}_{ab,cd} = \eta_{ad}\delta_b^{[e}\delta_c^{f]} + \eta_{bc}\delta_a^{[e}\delta_d^{f]} - \eta_{ac}\delta_b^{[e}\delta_d^{f]} - \eta_{bd}\delta_a^{[e}\delta_c^{f]} \,.$$

The notation with double indices is convenient since everything can be expressed in terms of the bilinear form

$$\eta_{ab} = \begin{bmatrix} \zeta & 0 & 0 & 0 \\ 0 & 1 & 0 & 0 \\ 0 & 0 & 1 & 0 \\ 0 & 0 & 0 & 1 \end{bmatrix} \,, \qquad \zeta = \pm 1 \,,$$

where the parameter $\zeta$ is equal to $+1$ or $-1$, for the cases of $SO(4)$ or $SO(3,1)$, respectively. The bilinear form $\eta_{ab}$ and its inverse $\eta^{ab}$ are used to raise and lower individual indices $a, b$, etc.

The $BF$ action can now be written as

$$S_{BF}[B, \omega] = \int_{M_4} B_{ab} \wedge R^{ab}(\omega) \,,$$

where the $\omega^{ab}$ is the spin connection 1-form, while $R^{ab}$ is the corresponding curvature 2-form. Note that in $D = 4$ the Lagrange multiplier $B^{ab}$ is a 2-form, and lends itself to constructing a 4-form $B^{ab} \wedge B^{cd}$, which can be used to deform the $BF$ action. It is precisely this type of deformation that gives us the Plebanski action:

$$\begin{aligned} S_{\text{Plebanski}}[B, \omega, \phi, \mu] = \int_{M_4} & B_{ab} \wedge R^{ab}(\omega) + \frac{1}{2}\phi_{abcd} \, B^{ab} \wedge B^{cd} \\ & + \mu \, \phi_{abcd} \left( a_1 \, g^{ab,cd} + a_2 \, \varepsilon^{abcd} \right) \,. \end{aligned} \tag{2.48}$$

The deformation term features a Lagrange multiplier 0-form $\phi_{abcd}$, which has the following symmetries,

$$\phi_{abcd} = -\phi_{bacd} = -\phi_{abdc} = +\phi_{cdab} \,,$$

then a Lagrange multiplier 4-form $\mu$, and two real parameters $a_1, a_2 \in \mathbb{R}$. Finally, there is an inverse Killing form

$$g^{ab,cd} = \frac{1}{2} \left( \eta^{ac} \eta^{bd} - \eta^{ad} \eta^{bc} \right) \,,$$

which satisfies the identities

$$g_{ab,cd}\, g^{cd,ef} = \delta^e_{[a} \delta^f_{b]} \,, \qquad g_{ab,cd}\, g^{ab,cd} = 6 \,, \qquad g^{ab,cd} \varepsilon_{abcd} = 0 \,.$$

Let us note here that original paper by Plebanski featured the special case $a_1 = 1$, $a_2 = 0$, while various more recent works instead commonly choose $a_1 = 0$, $a_2 = 1$, while others still choose to leave the two parameters arbitrary. For reasons of generality, we will do the same, and discuss special cases later on.

In order to see that the Plebanski action indeed reproduces general relativity, let us look at its equations of motion. Taking the variation with respect to $\phi_{abcd}$, $\mu$, $\omega^{ab}$ and $B^{ab}$, we obtain, respectively,

$$\frac{1}{2} B^{ab} \wedge B^{cd} + \mu \left( a_1 g^{ab,cd} + a_2 \varepsilon^{abcd} \right) = 0 \,, \tag{2.49}$$

$$\phi_{abcd} \left( a_1 g^{ab,cd} + a_2 \varepsilon^{abcd} \right) = 0 \,, \tag{2.50}$$

$$\nabla B^{ab} = 0 \,, \tag{2.51}$$

$$R_{ab} + \phi_{abcd} B^{cd} = 0 \,. \tag{2.52}$$

Equation (2.49) is called the *simplicity constraint*, and the deformation term in (2.48) is correspondingly called the *simplicity constraint term*, since one can demonstrate that it has a general solution of the form

$$B^{ab} = \frac{\alpha}{2} \varepsilon^{abcd} e_c \wedge e_d + \beta\, e^a \wedge e^b \,, \tag{2.53}$$

where $e^a$ are four 1-forms, identified with the tetrads, while the coefficients $\alpha$ and $\beta$ are related to the parameters $a_1$ and $a_2$ via

$$a_2 \alpha \beta = \frac{a_1}{4} \left( \alpha^2 + \zeta \beta^2 \right) \,.$$

In the special case when the parameter $a_1 = 0$ in the action, we have either $\alpha \neq 0$, $\beta = 0$, or $\alpha = 0$, $\beta \neq 0$ (the case $\alpha = \beta = 0$ is trivial), giving rise to the two families of solutions,

$$B^{ab} = \frac{\alpha}{2} \varepsilon^{abcd} e_c \wedge e_d \,, \quad \text{and} \quad B^{ab} = \beta\, e^a \wedge e^b \,,$$

both stating that the 2-forms $B^{ab}$ are simple,[4] hence the name simplicity constraint for (2.49). As for the remaining EoMs, equation (2.50) is called the *multiplier constraint*, along with the corresponding multiplier constraint term in (2.48). Finally, equation (2.51) can be recognized as the no-torsion equation when one substitutes (2.53), while after a certain amount of algebra (2.52) will likewise transform into the Einstein field equation. In this sense, Plebanski action (2.48) is classically equivalent to GR.

In the case of $T(M)$, the $BF$ term is given by (2.45), while

$$\int_M \phi_{abcd} B^{ab} \wedge B^{cd} \to \sum_{\sigma \in T(M)} \phi_{abcd}^v \sum_{\Delta, \Delta' \in \sigma} B^{ab}_\Delta B^{cd}_{\Delta'} \,,$$

and

$$\int_M \mu\, \phi_{abcd} \left(a_1 g^{ab,cd} + a_2 \varepsilon^{abcd}\right) \to \sum_{v \in T^*(M)} \mu^v \left(a_1 g^{ab,cd} + a_2 \varepsilon^{abcd}\right) \phi_{abcd}^v \,,$$

where $v$ is the dual vertex inside a 4-simplex $\sigma$.

### 2.4.3  *MacDowell-Mansouri action*

Let us consider an $SO(5 - k, k)$ Lie algebra for $k = 1, 2$. It can be written as

$$[J_{AB}, J_{CD}] = \eta_{[A|[C} J_{D]|B]} \,,$$

where $\eta = diag(1, 1, 1, 1, -1)$ for $k = 1$, or $\eta = diag(1, 1, 1, -1, -1)$ for $k = 2$, while $J_{AB} = -J_{BA}$, where $A, B = 1, 2, 3, 4, 5$. We also use

$$X_{[A} Y_{B]} \equiv X_A Y_B - X_B Y_A \,.$$

The group $SO(4, 1)$ is called de Sitter (dS) group, while the group $SO(3, 2)$ is called anti de Sitter (AdS) group. The AdS group is also the conformal group in the 3-dimensional Minkowski space.

---

[4]A 2-form $B$ is said to be *simple* (or *decomposable*) if it can be written as a wedge product of two 1-forms, say $B = a \wedge b$, where $a$ and $b$ are 1-forms. In $D \geq 4$ not every 2-form is simple. For example, $B$ can be a sum of two different simple 2-forms, $B = a \wedge b + c \wedge d$, where $a, b, c, d$ are different 1-forms. One can verify that simple 2-forms satisfy the equation $B \wedge B = 0$, while non-simple 2-forms do not.

In general, one defines a *rank* of a 2-form as the minimal number of simple 2-forms needed to express it. A 2-form $B$ has rank $p$ if and only if

$$\underbrace{B \wedge \cdots \wedge B}_{p} \neq 0 \qquad \text{and} \qquad \underbrace{B \wedge \cdots \wedge B}_{p+1} = 0 \,.$$

Simple 2-forms have rank $p = 1$.

Let us write the $SO(5-k,k)$ Lie algebra in an $SO(3,1)$ basis as

$$J_{AB} = (J_{ab}, J_{5a}) \equiv (J_{ab}, P_a)\,, \, a = 1, 2, 3, 4\,.$$

Consequently

$$[J_{ab}, J_{cd}] = \eta_{[a|[c}J_{d]|b]}\,,$$

$$[J_{ab}, P_c] = \eta_{[a|c}P_{|b]}\,,$$

$$[P_a, P_b] = \pm J_{ab}\,,$$

where the plus sign is in the dS case, while the minus sign is in the AdS case.

We can now introduce an $SO(5-k,k)$ connection

$$\mathcal{A} = \mathcal{A}^{AB}J_{AB} = \omega^{ab}J_{ab} + l^{-1}e^a P_a\,,$$

where $\omega^{ab}$ is the spin connection and $e_a$ are the tetrads on $M$. The parameter $l$ has a dimension of a length. The corresponding curvature is given by

$$\mathcal{F} = d\mathcal{A} + \mathcal{A} \wedge \mathcal{A} = F^{AB}J_{AB} = (R^{ab} + l^{-2}e^a \wedge e^b)J_{ab} + l^{-1}T^a P_a\,,$$

where $R^{ab} = d\omega^{ab} + \omega^a{}_c \wedge \omega^{cb}$ and $T^a = de^a + \omega^a{}_b \wedge e^b$ are the spacetime curvature and the torsion 2-forms.

The MacDowell-Mansouri (MDM) action [73] can be written as the $SO(5-k,k)$ $BF$ theory with a potential term which breaks the $SO(5-k,k)$ gauge symmetry down to the $SO(3,1)$ gauge symmetry, i.e.

$$S_{MDM} = \int_M \left(B^{AB} \wedge F_{AB} + \xi\,\varepsilon_{abcd}B^{ab} \wedge B^{cd}\right)\,.$$

The explicit $SO(3,1)$ symmetry invariant form of the action is given by

$$S_{MDM} = \int_M \left(B^{ab} \wedge (R_{ab} \pm l^{-2}e_a \wedge e_b) + l^{-1}\,b^a \wedge T_a + \xi\,\varepsilon_{abcd}B^{ab} \wedge B^{cd}\right)\,.$$

This action can be shown to be dynamically equivalent to

$$\tilde{S}_{MDM} = \xi^{-1}\int_M \varepsilon^{abcd}(R_{ab} \pm l^{-2}e_a \wedge e_b) \wedge (R_{cd} \pm l^{-2}e_c \wedge e_d)\,,$$

so that

$$\begin{aligned}
\tilde{S}_{MDM} = &\frac{1}{\xi}\int_M \varepsilon^{abcd}R_{ab} \wedge R_{cd} \\
&\pm \frac{2}{l^2\xi}\int_M \varepsilon^{abcd}R_{ab} \wedge e_c \wedge e_d \\
&+ \frac{1}{l^4\xi}\int_M \varepsilon^{abcd}e_a \wedge e_b \wedge e_c \wedge e_d\,.
\end{aligned} \tag{2.54}$$

The first term in (2.54) is a topological invariant, so that the dynamics is affected only by the last two terms, which are the Einstein-Hilbert action in terms of the Cartan variables[5] and the cosmological constant (CC) term. Consequently we have

$$G_N^{-1} \propto \frac{2}{l^2 \xi}, \quad \Lambda G_N^{-1} \propto \pm \frac{1}{l^4 \xi}. \qquad (2.55)$$

Therefore in the de Sitter group case we have a positive CC, while in the AdS case we have a negative CC.

The MDM action on $T(M)$ is given by

$$S_{MdM}^{T(M)} = \sum_{\Delta \in T(M)} B_\Delta^{AB} F_{AB}^f + \xi \sum_{\sigma \in T(M)} \sum_{\Delta, \Delta' \in \sigma} \varepsilon_{abcd} B_\Delta^{ab} B_{\Delta'}^{cd}.$$

## 2.5 *BFCG* theory

One way to generalize the notion of a group is to use the category theory, see [8] for a review and references. A category consists of objects and maps between the objects (called morphisms) such that natural composition rules between the morphisms are satisfied. A 2-category consists of objects, morphisms and maps between morphisms (called 2-morphisms) such that natural composition rules are satisfied.

A group is then a category with one object where all morphisms are invertible. Similarly, a 2-group is a 2-category with one object where all morphisms are invertible. This abstract definition leads to a concrete realization of a 2-group which is given by a crossed module $(G, H, \partial, \triangleright)$, see Appendix A for details.

A crossed module is a pair of groups $G$ and $H$, such that $\partial : H \to G$ is a homomorphism and $\triangleright$ is an action of $G$ on $H$ such that certain properties are satisfied, which are direct consequences of the categorical structure, see [8]. The elements of $G$ represent the 1-morphisms, while the elements of the semidirect product $G \times_s H$ represent the 2-morphisms. The canonical example of a 2-group relevant for physics is the Poincaré 2-group, where $G = SO(1, 3)$, $H = \mathbb{R}^4$, $\partial$ is a trivial homomorphism and $\triangleright$ is the usual action of the Lorentz transformations on the $\mathbb{R}^4$ space. The Lorentz group is the group of morphisms, while the usual Poincaré group is the group of 2-morphisms.

---

[5]The Cartan formulation of a manifold geometry uses the tetrads $e^a$ and the Lorentz group (spin) connection $\omega^{ab}$. For Riemannian geometries, the torsion $T^a = de^a + \omega^a{}_b \wedge e^b$ is zero. Some authors call the action $\int_M \varepsilon^{abcd} e_a \wedge e_b \wedge R_{cd}$ Palatini action, since Palatini used a formulation of a manifold geometry where the metric $g_{\mu\nu}$ and the affine connection $\Gamma^\lambda{}_{\mu\nu}$ are considered as independent variables.

One can construct a gauge theory on a 4-manifold $M$ based on a crossed module $(G, H, \partial, \triangleright)$ of Lie groups by using 1-forms $A$, which take values in the Lie algebra $\mathfrak{g}$ of $G$, and 2-forms $\beta$, which take values in the Lie algebra $\mathfrak{h}$ of $H$ [53, 82]. The forms $A$ and $\beta$ transform under the usual gauge transformations $g : M \to G$ as

$$A \to g^{-1}Ag + g^{-1}dg, \quad \beta \to g^{-1} \triangleright \beta, \tag{2.56}$$

while the gauge transformations generated by $H$ are given by

$$A \to A + \partial\eta, \quad \beta \to \beta + d\eta + A \wedge^{\triangleright} \eta + \eta \wedge \eta, \tag{2.57}$$

where $\eta$ is a 1-form taking values in $\mathfrak{h}$, see [82]. When the group $H$ is Abelian, which happens in the Poincaré 2-group case, then the $\eta \wedge \eta$ term in (2.57) vanishes, and one obtains the gauge transformations given in [53].

The pair $(A, \beta)$ represents a 2-connection on a 2-fiber bundle associated to the 2-Lie group $(G, H)$ and the manifold $M$. The corresponding curvature forms are given by

$$\mathcal{F} = dA + A \wedge A - \partial\beta, \quad \mathcal{G} = d\beta + A \wedge^{\triangleright} \beta, \tag{2.58}$$

and they transform as

$$\mathcal{F} \mapsto g^{-1}\mathcal{F}g, \quad \mathcal{G} \to g^{-1} \triangleright \mathcal{G}, \tag{2.59}$$

under the usual gauge transformations, while

$$\mathcal{F} \to \mathcal{F}, \quad \mathcal{G} \to \mathcal{G} + \mathcal{F} \wedge^{\triangleright} \eta, \tag{2.60}$$

under the $H$-gauge transformations.

One can introduce a natural topological gauge theory determined by the vanishing of the 2-curvature

$$\mathcal{F} = 0, \quad \mathcal{G} = 0. \tag{2.61}$$

These equations can be obtained from the action

$$S_{BFCG} = \int_M \langle B \wedge \mathcal{F} \rangle_{\mathfrak{g}} + \langle C \wedge \mathcal{G} \rangle_{\mathfrak{h}}, \tag{2.62}$$

where $B$ is a 2-form taking values in $\mathfrak{g}$, $C$ is a 1-form taking values in $\mathfrak{h}$, $\langle {-}, {-} \rangle_{\mathfrak{g}}$ is a $G$-invariant non-degenerate bilinear form in $\mathfrak{g}$ and $\langle {-}, {-} \rangle_{\mathfrak{h}}$ is a $G$-invariant non-degenerate bilinear form in $\mathfrak{h}$. The action (2.62) is called $BFCG$ action, in analogy with the action of $BF$ theory. The gauge transformations of the Lagrange multiplier fields are given by

$$B \to g^{-1}Bg, \quad C \mapsto g^{-1} \triangleright C, \tag{2.63}$$

for the usual gauge transformations, while

$$B \to B - [C, \eta], \quad C \mapsto C, \qquad (2.64)$$

for the $H$-gauge transformations.

The $BFCG$ action on $T(M)$ is given by

$$S_{BFCG}^{T(M)} = \sum_{\Delta \in T(M)} B_\Delta F_f + \sum_{\epsilon \in T(M)} C_\epsilon G_p,$$

where

$$C_\epsilon = \int_\epsilon C, \quad G_p = \int_p G,$$

and $p$ is a dual triangulation polyhedron which corresponds to an edge $\epsilon$, i.e. the vertices of $p$ are inside the 4-simplices which share the edge $\epsilon$.

### 2.5.1 *Poincaré 2-group and GR*

Let us now examine the case of the Poincaré 2-group. In this case $A = \omega^{ab} J_{ab}$, $\beta = \beta^a P_a$, where $a, b \in \{0, 1, 2, 3\}$, $J_{ab}$ are the generators of the Lorentz group while $P_a$ are the generators of the translation group $\mathbb{R}^4$. Consequently

$$\mathcal{F} = (d\omega^{ab} + \omega^a{}_c \wedge \omega^{cb}) J_{ab} = R^{ab} J_{ab}, \quad \mathcal{G} = (d\beta^a + \omega^a{}_b \wedge \beta^b) P_a = \nabla \beta^a P_a. \qquad (2.65)$$

The $G$-gauge transformations are the local Lorentz rotations

$$\omega \to g^{-1} \omega g + g^{-1} dg, \quad \beta \to g^{-1} \triangleright \beta, \qquad (2.66)$$

while the $H$-gauge transformations are the local translations

$$\delta_\varepsilon \omega = 0, \quad \delta_\varepsilon \beta^a = d\varepsilon^a + \omega^a{}_b \wedge \varepsilon^b, \qquad (2.67)$$

where $\eta = \varepsilon^a P_a$.

The $BFCG$ action then becomes

$$S_0 = \int_M (B^{ab} \wedge R_{ab} + C_a \wedge \nabla \beta^a), \qquad (2.68)$$

where

$$\delta_\varepsilon B = 0, \quad \delta_\varepsilon C = 0. \qquad (2.69)$$

Note that the transformation properties of the 1-forms $C^a$ are the same as the transformation properties of the tetrad 1-forms $e^a$ under the local Lorentz and the diffeomorphism transformations. Hence one can identify the $C$ fields with the tetrads and we write

$$S_0 = \int_M (B^{ab} \wedge R_{ab} + e^a \wedge \nabla \beta_a). \qquad (2.70)$$

The action (2.70) gives a theory of flat metrics, since $R^{ab} = 0$ implies the vanishing of the Riemann tensor. In order to obtain GR, we need that only the Ricci tensor vanishes. In the $BF$ theory approach to GR, this problem is resolved by constraining the $B$ field, such that $B^{ab} = \varepsilon^{abcd} e_c \wedge e_d$. Since in the 2-group formulation the tetrads are explicitly present, the required constraint is simply

$$B_{ab} = \varepsilon^{abcd} e_c \wedge e_d, \qquad (2.71)$$

or

$$*B_{ab} = e_a \wedge e_b. \qquad (2.72)$$

Hence the action for GR in the 2-group approach is given by

$$S = \int_M \left( B^{ab} \wedge R_{ab} + e^a \wedge \nabla \beta_a - \phi_{ab} \wedge (B^{ab} - \varepsilon^{abcd} e_c \wedge e_d) \right). \qquad (2.73)$$

The equations of motion are

$$R_{ab} - \phi_{ab} = 0, \qquad (2.74)$$

$$\nabla \beta_a + 2\varepsilon_{abcd} \phi^{bc} \wedge e^d = 0, \qquad (2.75)$$

$$\nabla B_{ab} - e_{[a} \wedge \beta_{b]} = 0, \qquad (2.76)$$

$$\nabla e_a = 0, \qquad (2.77)$$

$$B_{ab} - \varepsilon_{abcd} e^c \wedge e^d = 0, \qquad (2.78)$$

which are obtained by varying $S$ with respect to $B$, $e$, $\omega$, $\beta$ and $\phi$, respectively.

From $B = *(e \wedge e)$ it follows that $\nabla B \propto *(e \wedge \nabla e)$, so that $\nabla B = 0$ due to (2.77). The equation (2.76) then implies that $e_{[a} \wedge \beta_{b]} = 0$. For invertible tetrads we then obtain $\beta = 0$, see Appendix B for proof. Therefore (2.74) and (2.75) imply

$$\varepsilon_{abcd} R^{bc} \wedge e^d = 0. \qquad (2.79)$$

The equation (2.79) is the same as the equation of motion for the EH action in the Cartan formalism

$$S_{EHC} = \int_M \varepsilon_{abcd} e^a \wedge e^b \wedge R^{cd}, \qquad (2.80)$$

for the $e$ variations, while (2.77) is equivalent to $\delta S_{EHC}/\delta \omega$ when the tetrads are invertible.

### 2.5.2 Coupling of matter in BFCG GR

Since the tetrads are present in the $BFCG$ action, the coupling of matter fields is essentially given by the coupling of matter fields in the EC formulation. The only subtlety is in the coupling of fermions, since their presence introduces a non-zero torsion.

The Dirac action for the fermion field in the EC formulation is given by

$$S_D = i\kappa_1 \int \varepsilon_{abcd} e^a \wedge e^b \wedge e^c \wedge \bar{\psi} \left( \gamma^d \overset{\leftrightarrow}{d} + \{\omega, \gamma^d\} + \frac{im}{2} e^d \right) \psi, \quad (2.81)$$

where $\omega = \omega_{ab}[\gamma^a, \gamma^b]/8$ and $\kappa_1 = 8\pi l_p^2/3$. The $\delta(S_{EC} + S_D)/\delta\omega$ equation gives the torsion $T_a \equiv \nabla e_a = -\kappa_2 s_a$, where

$$s_a = i\varepsilon_{abcd} e^b \wedge e^c \, \bar{\psi}\gamma_5\gamma^d\psi,$$

is the spin 2-form, and $\kappa_2 = -3\kappa_1/4$. Hence in the $BFCG$ formulation we need a term $\int_M \beta_a \wedge s^a$ in the action.

Let us consider the action

$$S_m = S + S_D + S_{\beta\psi}, \quad (2.82)$$

where

$$S_{\beta\psi} = i\kappa_2 \int \varepsilon_{abcd} e^a \wedge e^b \wedge \beta^c \, \bar{\psi}\gamma_5\gamma^d\psi.$$

By varying $S_m$ with respect to $B$, $e$, $\omega$, $\beta$, $\phi$ and $\bar{\psi}$, respectively, we obtain

$$R_{ab} - \phi_{ab} = 0, \quad (2.83)$$

$$\nabla\beta_a + \varepsilon_{abcd} e^b \wedge \left[ 2\phi^{cd} - \frac{3i\kappa_1}{2}\beta^c\bar{\psi}\gamma_5\gamma^d\psi \right.$$
$$\left. + 3i\kappa_1 e^c \wedge \bar{\psi}\left( \gamma^d \overset{\rightarrow}{\nabla} - \overset{\leftarrow}{\nabla}\gamma^d + \frac{im}{6}e^d \right)\psi \right] = 0, \quad (2.84)$$

$$\nabla B_{ab} - e_{[a} \wedge \beta_{b]} - 2\kappa_2\varepsilon_{abcd}e^c \wedge s^d = 0, \quad (2.85)$$

$$\nabla e_a + \kappa_2 s_a = 0, \quad (2.86)$$

$$B_{ab} - \varepsilon_{abcd}e^c \wedge e^d = 0, \quad (2.87)$$

$$i\kappa_1\varepsilon_{abcd}e^a \wedge e^b \wedge \left( 2e^c \wedge \gamma^d\nabla + \frac{im}{2}e^c \wedge e^d - 3(\nabla e^c)\gamma^d - \frac{3}{4}\beta^c\gamma_5\gamma^d \right)\psi = 0. \quad (2.88)$$

Exactly like in the pure gravity case, from (2.87) it follows that $\nabla B \propto *(e \wedge \nabla e)$, so that (2.85) gives

$$2\,\varepsilon_{abcd}e^c \wedge \left(\nabla e^d + \kappa_2 s^d\right) + e_{[a} \wedge \beta_{b]} = 0\,.$$

This equation gives $e_{[a} \wedge \beta_{b]} = 0$, due to (2.86). If the tetrads are invertible, one then obtains $\beta^a = 0$, so that (2.84) gives

$$\varepsilon^{abcd}e_b \wedge \left(R_{cd} - T^{\psi}_{cd}\right) = 0\,, \tag{2.89}$$

where $T^{\psi}_{ab}$ is the energy-momentum 2-form for the fermions. The equation (2.89) is equivalent to the Einstein equations when a Dirac fermion is coupled to EC gravity.

The $\delta S_m/\delta\psi$ and $\delta S_m/\delta\bar{\psi}$ equations are related by the spinor conjugation. For the invertible tetrads, and by using $\nabla e = -\kappa_2 s$, the equation (2.88) reduces to the usual Dirac equation

$$(i\gamma^\mu \nabla_\mu - m)\,\psi = 0\,, \tag{2.90}$$

where $\gamma^\mu = e^\mu{}_a \gamma^a$.

As far as scalar and YM fields are concerned, they do not couple to $\omega$, so that one simply adds the corresponding EC formalism terms to $S_m$

$$S_m \to S_m + \int_M |e|\left(g^{\mu\nu}\partial_\mu\Phi\partial_\nu\Phi + g^{\mu\nu}g^{\rho\sigma}\,Tr\,F_{\mu\rho}F_{\nu\sigma}\right)d^4x\,, \tag{2.91}$$

where $g_{\mu\nu} = e^a_\mu e^b_\nu \eta_{ab}$.

One can also introduce the Immirzi parameter $\gamma$, by adding an additional term $S_\gamma$ to the action $S_m$, where

$$S_\gamma = -\frac{1}{\gamma}\int \phi^{ab} \wedge e_a \wedge e_b + \frac{i\kappa_2}{\gamma^2+1}\int \varepsilon_{abcd}e^a \wedge e^b \wedge \beta^c \bar{\psi}\gamma_5 \gamma^d \psi$$
$$+ \frac{i\kappa_2\gamma}{\gamma^2+1}\int e^a \wedge e^b \wedge \beta_a \bar{\psi}\gamma_5\gamma_b\psi\,.$$

The resulting equations of motion are equivalent to the equations of motion obtained from the action $S_{EC} + S_D + S_H$, where $S_H$ is the Holst term [62]

$$S_H = -\frac{2}{\gamma}\int e^a \wedge e^b \wedge R_{ab}\,.$$

The physical motivation for the introduction of the Immirzi parameter lies in the fact that it is the coupling constant between fermions and torsion, as discussed in detail in [50, 105].

As a final comment, note that it is also completely straightforward to add a cosmological constant term to the action, in the form

$$S_\Lambda = \Lambda \int \varepsilon_{abcd}\, e^a \wedge e^b \wedge e^c \wedge e^d\,.$$

# Chapter 3

# State-sum models of QG

## 3.1 *BF* theory path integral

Consider a *BF* theory for a compact Lie group $G$ on a closed compact 4-manifold $M$. One can define the path integral (PI) by using a heuristic expression

$$Z(M) = \int \mathcal{D}A \int \mathcal{D}B \exp\left(i \int_M Tr\, B \wedge F\right) = \int \mathcal{D}A\, \delta(F). \qquad (3.1)$$

One can define the PI (3.1) more precisely by using a triangulation $T(M)$. The idea behind this is to use the property of a *BF* theory that it does not have local DoF, so that one expects that the PI will be triangulation independent. If one can construct $Z(T(M))$ which is triangulation independent, then

$$Z(T(M)) = Z(M). \qquad (3.2)$$

Let $T^*(M)$ be a dual triangulation of $T(M)$. In a dual triangulation there is a dual vertex $v$ in the interior of each 4-simplex $\sigma$, an edge $l$ dual to a tetrahedron $\tau$ shared by two 4-simplices, a face $f$ dual to a triangle $\Delta$ shared by several 4-simplices, a polyhedron (3-complex) $p$ dual to an edge $\epsilon$ and a 4-complex $q$ dual to a triangulation vertex $\nu$ in its interior.

We can then define the simplex variables

$$A_l = \int_l A, \qquad F_f = \int_f F, \qquad B_\Delta = \int_\Delta B,$$

so that

$$\int_M Tr\, B \wedge F = \sum_\Delta \int_{D(f,\Delta)} Tr\, B \wedge F \approx \frac{1}{4} \sum_\Delta Tr\, B_\Delta\, F_f,$$

where $D(f, \Delta)$ is a diamond-like 4-complex spanned by a triangle $\Delta$ and the dual face $f$. We use a short-hand notation

$$\sum_s X_s = \sum_{s \in T} X_s,$$

where $s$ is a simplex, and we will also use

$$\prod_s X_s = \prod_{s \in T} X_s.$$

By integrating the $B_\Delta$ variables in the path integral

$$Z(M) = \int \prod_l dA_l \int \prod_\Delta dB_\Delta \exp\left(i \sum_\Delta Tr\, B_\Delta \wedge F_f\right),$$

we obtain

$$Z(M) = \int_{\mathfrak{g}^L} \prod_l dA_l \prod_f \delta(F_f). \tag{3.3}$$

The Lie-algebra integral (3.3) can be written as an integral over $G^L$, by using

$$g_l = e^{A_l}, \quad \delta(F_f) = \delta(g_f), \quad g_f = \prod_{l \in \partial f} g_l,$$

so that

$$Z(M) = \int_{G^L} \prod_l dg_l \prod_f \delta(g_f). \tag{3.4}$$

The group delta function $\delta(g)$ is defined by

$$\delta(g) = \sum_\Lambda \dim \Lambda\, \chi_\Lambda(g),$$

where $\Lambda$ are the irreducible representations (irreps) of $G$ and $\chi_\Lambda$ is the character function, i.e. the trace of the representation matrix $D^{(\Lambda)}(g)$.

The group integral (3.4) can be then written as a sum over the representations $\Lambda_f$ associated to the faces, i.e. as a state sum. We can perform the $g_l$ integrations by using

$$\int_G D^{(\Lambda_1)}_{\alpha_1 \beta_1}(g)\, D^{(\Lambda_2)}_{\alpha_2 \beta_2}(g)\, D^{(\Lambda_3)}_{\alpha_3 \beta_3}(g)\, D^{(\Lambda_4)}_{\alpha_4 \beta_4}(g)\, dg$$
$$= \sum_\iota C^{\Lambda_1 \Lambda_2 \Lambda_3 \Lambda_4(\iota)}_{\alpha_1 \alpha_2 \alpha_3 \alpha_4}\, \bar{C}^{\Lambda_1 \Lambda_2 \Lambda_3 \Lambda_4(\iota)}_{\beta_1 \beta_2 \beta_3 \beta_4},$$

where $\iota \in Hom(\Lambda_1, \Lambda_2, \Lambda_3, \Lambda_4)$ counts the intertwiners, since each dual edge $l$ is shared by four faces. Consequently we obtain for $T^*(M)$

$$Z = \sum_{\Lambda, \iota} \prod_f \dim \Lambda_f \prod_v A_v(\Lambda, \iota), \tag{3.5}$$

or equivalently for $T(M)$

$$Z = \sum_{\Lambda, \iota} \prod_{\Delta} \dim \Lambda_{\Delta} \prod_{\sigma} A_{\sigma}(\Lambda, \iota). \qquad (3.6)$$

The sum in (3.6) is over the labels of the amplitudes for a labeled 2-complex (which is called a spin foam). The vertex amplitude $A_v(\Lambda, \iota)$, or the 4-simplex amplitude $A_{\sigma}$, is given by the product of five $C^{(\iota)}$ tensors, whose representation indices are contracted by $\delta_{\alpha\alpha'}$, so that $A_{\sigma}$ is a function of ten $\Lambda_{\Delta}$ and five $\iota_{\tau}$ which are associated with a 4-simplex $\sigma$. In the case of the $SU(2)$ group, this gives rise to a $15j$-symbol, see [102].

The advantage of representing the integral (3.4) as a dual 2-complex state sum (3.5) is that it is easier to analyze the triangulation independence. This is done by using the Pachner moves [103], which are local, i.e. they can be applied to one 4-simplex, or to two adjacent 4-simplices, or to three adjacent 4-simplices. The terms in the sum (3.5) are invariant under the Pachner moves,[1] so that the sum is formally triangulation independent. However, the sum (3.5) is infinite and it is not convergent, and this requires a regularization. If one wants to obtain a manifold invariant, this can be done by using a quantum group $U_q(\mathfrak{g})$ associated to a Lie group $G$, where $q$ is a root of unity, see [36].

In the case of a $BF$ theory in 3 dimensions, by using the same method, one obtains the same expression as (3.5), but now a dual face corresponds to an edge and a vertex amplitude $A_v$ corresponds to a tetrahedron. $A_v$ is then given as the contracted product of four Clebsch-Gordon (CG) coefficients. In the case of the $SU(2)$ group, $A_v$ is known as the $6j$-symbol, and Ponzano and Regge were the first to formulate the corresponding (PR) state sum [108],

$$Z_{PR} = \sum_{j} \prod_{\epsilon} (2j_{\epsilon} + 1) \prod_{\tau} \{6j\}_{\tau}. \qquad (3.7)$$

Note that the intertwiners have disappeared from the PR state sum because the CG coefficients for $SU(2)$ irreps have trivial intertwiners. The quantum group regularization of the PR state sum is known as the Turaev-Viro invariant for a 3-manifold [121].

## 3.2 Spin-foam models of QG

The classical dynamics of the EH action can be recovered by using the Plebanski action (2.48), which is a constrained $BF$ theory action for the

---

[1] Up to some sign factors, see the $q \to 1$ limit of [36].

Lorentz group $SO(3,1)$. Since the path-integral quantization of a $BF$ theory leads to a state-sum model, the main idea is to use the $BF$ theory state sum (3.6) and to impose the Plebanski constraints by restricting the set of representation one is using to color the spin foam.

Hence a face from a spin foam (SF) should carry a Lorentz group irrep $\Lambda$ from some suitable chosen subset of the Lorentz group irreps. Since an edge of a SF face, carrying a representation $\Lambda$, can belong to a boundary spin network, then the label $\Lambda$ should be equivalent to an $SU(2)$ irrep label $j$. Furthermore, the SF irreps should satisfy the constraints which follow from the Plebanski action when restricted to a tetrahedron in the SF 2-complex, see [104].

Another restriction is that one must take the set of unitary irreps of $SO(3,1)$, so that $\Lambda = (k, \rho)$, where $\kappa \in \mathbb{Z}$ and $\rho \in \mathbb{R}$. The reason is that the finite-dimensional non-unitary irreps $\Lambda = (j_+, j_-)$, where $j_\pm \in \mathbb{Z}_+/2$, describe the same model as the Euclidean gravity SF model based on the $SO(4)$ group. Furthermore, the Plebanski constraints imply $k = 2j$ and $\rho = 2\gamma j$, where $j \in \mathbb{N}_0/2$.

A SF transition amplitude from an initial spin network $\hat{\gamma}_1$ and to a final spin network $\hat{\gamma}_2$ can be only defined on a triangulation of a manifold with globally hyperbolic topology, $M = \Sigma \times I$. Hence

$$A(\hat{\gamma}_1, \hat{\gamma}_2, T(M)) = \sum_{\hat{\Gamma}\,;\,T(\Gamma) \subset T(M)} c_{\hat{\Gamma}} A_{\hat{\Gamma}}(\hat{\gamma}_1, \hat{\gamma}_2)\,, \qquad (3.8)$$

where $T(\Gamma)$ is an immersion of the 2-complex $\Gamma$ into $T(M)$. Given the definition (3.8), one has to solve three problems:

(1) find a definition of $A_{\hat{\Gamma}}$ which has the correct semiclassical limit,
(2) perform the sum over the spin foams, and
(3) construct the smooth limit $T(M) \to M$.

As we discussed earlier, the problems (2) and (3) are very difficult and very little progress has been made so far, while the problem (1) has been solved, see the next section.

### 3.2.1  *Semiclassical limit of a SF vertex amplitude*

The semiclassical (SC) limit of $A_{\hat{\Gamma}}$ is defined in the following way. Let the 2-complex $\Gamma$ be a subcomplex of the dual 2-complex of $T(M)$, so that $j_f = j_\Delta$ and $\iota_l = \iota_\tau$, and let $j_\Delta \to \infty$. If we have the asymptotics

$$A_\Gamma(j, \iota) \approx \begin{cases} N_\Gamma(j, \iota)\, e^{iS_\Gamma(j,\iota)/l_P^2} & (j, \iota) \text{ geometric}, \\ 0 & (j, \iota) \text{ non-geometric}, \end{cases} \qquad (3.9)$$

where $S_\Gamma(j, \iota) = S_R(L)$ and $S_R$ is the Regge action, we can say that we have the correct semiclassical limit.

Note that for a geometric $(j, \iota)$ configuration we can express $j_\Delta$ and $\iota_\tau$ as functions of the edge lengths $L_\epsilon$ and if $S_\Gamma(j, \iota) = S_R(L)$ it may happen that

$$\sum_{j, \iota} A_\Gamma(j, \iota) \approx \int \prod_\epsilon dL_\epsilon \, \mu(L) \, e^{iS_R(L)/l_P^2} .$$

This then indicates that the classical limit of such a spin foam model is the Regge action. In chapter 4 we shall see how to rigorously define the classical limit of a SF model and how to calculate the semiclassical effects by using the effective action formalism.

The association of the edge lengths $L_\epsilon$ to the SF labels $j_\Delta$ and $\iota_\tau$ comes from the fact that $\sqrt{j_\Delta(j_\Delta + 1)} l_P^2 \propto A_\Delta$ (area of a triangle) and that the geometric meaning of $\iota_\tau$ is related to the unit normals $\vec{n}_{\Delta,\tau}$, which are equivalent to the dihedral angles $\phi_{\epsilon,\tau}$ in a tetrahedron. Namely, there is a relation

$$\sum_{\iota_\tau} A_\sigma(j, \iota) = \int_{(S^2)^{20}} \prod_{\Delta,\tau} d^2 \vec{n}_{\Delta,\tau} W_\sigma(j, \vec{n}) ,$$

where

$$A_\Gamma(j, \iota) = \prod_{\Delta \in T(\Gamma)} \dim j_\Delta \prod_{\sigma \in T(\Gamma)} A_\sigma(j, \iota) ,$$

and $\vec{n}$ are the normal vectors for the triangles of a tetrahedron, see [72]. On the other hand,

$$\cos \phi_{\epsilon,\tau} = -\vec{n}_{\Delta,\tau} \cdot \vec{n}_{\Delta',\tau} , \quad \epsilon = \Delta \cap \Delta' .$$

and as we discussed in section 2.3, the metric geometry of a spin foam is determined by the areas $A_\Delta$ and the dihedral angles $\phi$, provided the following constraints are satisfied

$$A_\Delta = \sum_{\Delta' \in \tau} A_{\Delta'} \cos \phi_{\Delta' \cap \Delta, \tau} , \tag{3.10}$$

$$\cos \alpha_{p,\Delta}(\phi) = \cos \tilde{\alpha}_{p,\Delta}(\tilde{\phi}) , \tag{3.11}$$

where $\alpha_{p,\Delta}$ is the dihedral angle for a triangle $\Delta$ at a vertex $p$ expressed as a function of $\phi$'s for a tetrahedron $\tau$ which contains $\Delta$, while $\tilde{\alpha}$ is the same angle expressed as a function of $\phi$'s for the tetrahedron $\tilde{\tau}$ which also contains the triangle $\Delta$.

The constraint (3.10) is also known as the closure constraint, and it is equivalent to

$$\sum_{\Delta \in \tau} B_\Delta^{ab} = 0 \,, \tag{3.12}$$

where $B_\Delta^{ab}$ is the simplicial complex analog of the 2-form $B^{ab}$ from the Plebanski action. The constraint (3.12) is a simplicial complex analog of the Gauss constraint from canonical loop QG. The second set of constraints (3.11) can be implemented in the Plebanski formulation via the quadratic constraints

$$\eta_{ab,cd}^i B_\Delta^{ab} B_\Delta^{cd} = 0 \,, \quad i \in \{1,2\} \,, \tag{3.13}$$

where $\eta^1 = \varepsilon_{abcd}$ and $\eta^2 = \eta_{[a|[c}\eta_{d]|b]}$. These are the simplicial analogs of the Plebanski constraints, see section 2.4.2. In the case of the Holst action,[2] the constraints (3.13) are modified by substituting $B_{ab}$ with $B_{ab} + \gamma \varepsilon_{abcd} B^{cd}$. The quadratic constraints are then applied to the unitary Lorentz irreps $(k,\rho)$ by using a quantization prescription $B^{ab} = \hat{J}^{ab}$, which then gives $(k,\rho) = (2j, 2\gamma j)$, see [44].

However, the EPRL/FK implementation of the quadratic constraints via the ansatz $B_{ab} = \hat{J}_{ab}$ is difficult to understand at the classical level, so that one has to perform the semiclassical analysis of the amplitudes in order to see whether one obtains the correct asymptotics. As we will show, one does not obtain the asymptotics (3.9) exactly, since $S_\Gamma$ is given by the area-Regge action, see section 4.3.

## 3.3   The EPRL/FK model

The EPRL/FK spin-foam model state sum is given by

$$Z(T) = \sum_{j,\iota} \prod_{f \in T^*} A_2(j_f) \prod_{v \in T^*} W(j_{f(v)}, \iota_{e(v)}) \,, \tag{3.14}$$

where $T$ is a triangulation of the spacetime manifold, $T^*$ is the dual simplicial complex, while $e$, $f$ and $v$ denote the edges, the faces and the vertices of $T^*$, respectively. The sum in (3.14) is over all possible assignments of $SU(2)$ spins $j_f$ to the faces of $T^*$ (triangles of $T$) and over the corresponding intertwiner assignments $\iota_e$ to the edges of $T^*$ (tetrahedrons of $T$). $A_2$ is the face amplitude, and it can be fixed to be

$$A_2(j) = \dim j = 2j + 1 \,, \tag{3.15}$$

---

[2]The Holst action is the Einstein-Cartan action where $R_{ab} \to R_{ab} + \gamma \varepsilon_{abcd} R^{cd}$.

by using the consistent gluing requirements for the transition amplitudes between three-dimensional boundaries, see [20].

The vertex amplitude $W$ can be written as

$$W(j_f, \iota_e) = \sum_{k_e \geqslant 0} \int_0^{+\infty} d\rho_e \, (k_e^2 + \rho_e^2) \left( \bigotimes_e f_{k_e \rho_e}^{\iota_e}(j_f) \right)$$
$$\{15j\}_{SL(2,\mathbb{C})} \left( (2j_f, 2\gamma j_f); (k_e, \rho_e) \right), \qquad (3.16)$$

where the $15j$-symbol is for the unitary representations $(k, \rho)$ of the $SL(2, \mathbb{C})$ group, the universal covering group of the Lorentz group. The $f_{k_e \rho_e}^{\iota_e}$ are the fusion coefficients, defined in detail in [34, 44, 49].

Instead of using the spin-intertwiner basis, one can rewrite (3.14) in the coherent state basis, introduced in [72]. In this basis, the state sum is given by

$$Z(T) = \sum_j \int \prod_{e,f} d^2 \vec{n}_{ef} \prod_{f \in T^*} \dim j_f \prod_{v \in T^*} W(j_{f(v)}, \vec{n}_{e(v)f(v)}). \qquad (3.17)$$

The $\vec{n}_{ef}$ is a unit three-dimensional vector associated to the triangle dual to a face $f$ of the tetrahedron dual to an edge $e$ which belongs to $f$ (see [72] for details). For a geometric tetrahedron, the four vectors $\vec{n}$ can be identified with the unit normal vectors for the triangles. Note that the domain of integration for each such vector is a 2-sphere.

The key property of $W(j, \vec{n})$ amplitude, which was used to find the large-spin asymptotics, is that it can be written as an integral over the manifold $SL(2, \mathbb{C})^4 \times (\mathbb{CP}^1)^{10}$, see [13]. More precisely,

$$W(j, \vec{n}) = \text{const} \cdot \prod_{k=1}^{10} \dim j_k \int_{SL(2,\mathbb{C})^5} \prod_{a=1}^{5} dg_a \delta(g_5)$$
$$\int_{(\mathbb{CP}^1)^{10}} \prod_{k=1}^{10} dz_k \, \Omega(g, z) \, e^{S(j, \vec{n}, g, z)}, \qquad (3.18)$$

where $\Omega$ is a slowly changing function and

$$S(j, \vec{n}, g, z) = \sum_{k=1}^{10} j_k \log w_k(\vec{n}, g, z) = \sum_{k=1}^{10} j_k \left( \ln |w_k(\vec{n}, g, z)| + i\theta_k(\vec{n}, g, z) \right).$$

The functions $w_k$ are complex-valued, so that $\theta_k = \arg w_k + 2\pi m_k$, where $m_k$ are integers which have to be chosen such that $\log w_k$ belong to the same branch of the logarithm. Since $|w_k| \leqslant 1$, it follows that $Re\, S \leqslant 0$ and it can be shown that the large-spin asymptotics is given by

$$W(\lambda j, \vec{n}) \approx \frac{\text{const}}{\lambda^{12}} \sum_{x^*} \frac{\Omega(x^*) \, e^{i\lambda \sum_k j_k \theta_k(\vec{n}, x^*)}}{\sqrt{\det(-H(j, \vec{n}, x^*))}}, \qquad (3.19)$$

for $\lambda \to +\infty$, where the sum is over the critical points $x^* = (g^*, z^*)$ satisfying

$$Re\, S(j, \vec{n}, g^*, z^*) = 0\,, \quad \left.\frac{\partial S}{\partial g_a}\right|_{x^*} = 0\,, \quad \left.\frac{\partial S}{\partial z_k}\right|_{x^*} = 0\,, \tag{3.20}$$

and $H(x)$ is the Hessian for the function $S(x)$, see [13].

There are finitely many critical points, and it can be shown that the conditions (3.20) require that $j_k$ are proportional to the areas of triangles for a geometric 4-simplex, while $\vec{n}$ have to be the normal vectors for the triangles in a tetrahedron of a geometric 4-simplex and $g^*$ have to be the corresponding holonomies. A geometric 4-simplex has a consistent assignment of the edge-lengths, and it can be shown that $\theta_k(\vec{n}, x^*)$ is proportional to the dihedral angle for a triangle in a geometric 4-simplex, so that

$$S_R^{(v)} = \sum_{k=1}^{10} j_k\, \theta_k(\vec{n}, x^*)$$

corresponds to the Regge action for a 4-simplex.

The Hessian $H(j, \vec{n}, x)$ is a $44 \times 44$ matrix, and

$$H_{\alpha\beta}(j, \vec{n}, x^*) = \sum_{k=1}^{10} j_k\, H_{\alpha\beta}^{(k)}(\vec{n}, x^*)\,, \tag{3.21}$$

since $S$ is a linear function of $j$. Consequently

$$\det(-H) = \sum_{m_1 + \cdots + m_{10} = 44} (j_1)^{m_1} \cdots (j_{10})^{m_{10}} D_{m_1 \ldots m_{10}}(\vec{n}, x^*)\,, \tag{3.22}$$

is a homogeneous polynomial of degree 44 in $j_k$ variables. One also has that $Re\,(-H)$ is a positive definite matrix.

### 3.3.1 A bound for the vertex amplitude

We will now find a bound for the vertex amplitude by using the asymptotic formula (3.19) and its generalization for the case when some of the vertex spins are large and others are small. Since $\lambda S(j, \vec{n}, x) = S(\lambda j, \vec{n}, x)$ and

$$\lambda^{44} \det(-H(j, \vec{n}, x^*)) = \det(-H(\lambda j, \vec{n}, x^*))\,,$$

then the formula (3.19) can be rewritten as

$$W(j, \vec{n}) \approx const \prod_{k=1}^{10} \dim j_k \sum_{x^*} \frac{\Omega(x^*)\, e^{i \sum_k j_k \theta_k(\vec{n}, x^*)}}{\sqrt{\det(-H(j, \vec{n}, x^*))}}\,,$$

when $j = (j_1, \ldots, j_{10}) \to (+\infty, \ldots, +\infty) \equiv (+\infty)^{10}$, because $\prod_{k=1}^{10} \dim j_k$ scales as $\lambda^{10}$ for $\lambda$ large. Therefore

$$\lim_{j \to (+\infty)^{10}} W(j, n) = \text{const} \lim_{j \to (+\infty)^{10}} \prod_{k=1}^{10} \dim j_k \sum_{x^*} \frac{\Omega(x^*) \, e^{i \sum_k j_k \theta_k (\vec{n}, x^*)}}{\sqrt{\det(-H(j, \vec{n}, x^*))}} \,. \tag{3.23}$$

Note that

$$\left| \sum_{x^*} \frac{\Omega(x^*) \, e^{i \sum_k j_k \theta_k (\vec{n}, x^*)}}{\sqrt{\det(-H(j, \vec{n}, x^*))}} \right| \leqslant \sum_{x^*} \frac{|\Omega(x^*)|}{\sqrt{|\det(-H(j, \vec{n}, g^*))|}} \,, \tag{3.24}$$

and

$$\lim_{j \to (+\infty)^{10}} \frac{\prod_{k=1}^{10} \dim j_k}{\sqrt{|\det(-H(j, \vec{n}, x^*))|}} = 0 \,, \tag{3.25}$$

due to (3.22). The equations (3.23), (3.24) and (3.25) imply

$$\lim_{j \to (+\infty)^{10}} W(j, \vec{n}) = 0 \,. \tag{3.26}$$

The equation (3.26) is equivalent to

$$\forall \epsilon > 0 \,, \, \exists \delta > 0 \quad \text{such that} \quad j_1 > \delta \,, \ldots, \, j_{10} > \delta \Rightarrow |W(j, \vec{n})| < \epsilon \,.$$

This implies that $W$ is a bounded function in the region

$$D_{10} = \{j \,|\, j_1 > \delta \,, \ldots, \, j_{10} > \delta\} \,.$$

If we denote with $D_m$ the region where $m < 10$ spins are greater than $\delta$ and the rest are smaller or equal than $\delta$, then

$$\mathbb{R}_+^{10} \setminus D_{10} = \bigcup_{m=0}^{9} D_m \,.$$

Since the regions $D_m$ are not compact for $m > 0$, we do not know whether $W$ is bounded in these regions. In order to determine this we need to know the asymptotics of $W$ for the cases when some of the spins are large and others are small. This asymptotics can be obtained by using the same method as in the case when all the vertex spins are large.

Let $m$ be the number of large spins ($m \geqslant 3$ due to the triangle inequalities for the vertex spins) and let $j' = (j_1, \ldots, j_m)$ and $j'' = (j_{m+1}, \ldots, j_{10})$. Then

$$S(\lambda j', j'', n, x) = \sum_{k=1}^{m} \lambda j_k' (\ln |w_k| + i \theta_k) + \sum_{k=m+1}^{10} j_k'' (\ln |w_k| + i \theta_k)$$

$$= \lambda S_m(j', n, x) + O(1) \,.$$

Therefore the asymptotic properties of $W(j', j'', n)$ will be determined by the critical points of $S_m(j', n, x)$. Consequently

$$W(\lambda j', j'', \vec{n}) \approx \frac{\text{const}}{\lambda^{r/2-m}} \sum_{x^*} \frac{\Omega(x^*) e^{i\lambda \sum_{k=1}^m j'_k \theta_k(\vec{n}, x^*)}}{\sqrt{\det(-\tilde{H}_m(j', \vec{n}, x^*))}}, \tag{3.27}$$

where $r$ is the rank of the Hessian matrix $H_m$ for $S_m$ at a critical point $x^*$ ($1 \leqslant r \leqslant 44$) and $\tilde{H}_m$ is the reduced Hessian matrix. $\tilde{H}_m$ is the restriction of the Hessian $H_m$ to the orthogonal complement of $Ker\, H_m$ and $\tilde{H}_m$ has to be used if $r < 44$.

The asymptotics (3.27) implies that the function $W(j', j'', \vec{n})$ will vanish for large $j'$ if $r/2 - m > 0$. If this was true for all $m$ we could say that $W(j)$ is a bounded function in $\mathbb{R}_+^{10}$. However, calculating the values for $r$ is not easy. Instead, we are going to estimate $|W(j', j'', \vec{n})|$. Note that (3.27) is equivalent to

$$W(j', j'', \vec{n}) \approx \text{const} \prod_{k=1}^m \dim j'_k \sum_{x^*} \frac{\Omega(x^*) e^{i\lambda \sum_{k=1}^m j'_k \theta_k(\vec{n}, x^*)}}{\sqrt{\det(-\tilde{H}_m(j', \vec{n}, x^*))}}$$

for $j' \to (+\infty)^m$, since $S_m$ and $\tilde{H}_m$ are linear functions of the spins $j'$ and $\det(-\tilde{H}_m)$ scales as $\lambda^r$, while $\prod_{k=1}^m \dim j_k$ scales as $\lambda^m$ when $j' \to \lambda j'$ and $\lambda$ is large. Hence

$$\frac{W(j', j'', \vec{n})}{\prod_{k=1}^m \dim j'_k} \approx \text{const} \sum_{x^*} \frac{\Omega(x^*) e^{i\lambda \sum_{k=1}^m j'_k \theta_k(\vec{n}, x^*)}}{\sqrt{\det(-\tilde{H}_m(j', \vec{n}, x^*))}},$$

for $j' \to (+\infty)^m$. From here it follows that for every $m \geqslant 3$

$$\lim_{j \to (+\infty)^m} \frac{W(j', j'', \vec{n})}{\prod_{k=1}^m \dim j'_k} = 0,$$

since $r(m) \geqslant 1$. Given that $W = 0$ in $D_1$ and $D_2$, it follows that

$$\frac{W(j, \vec{n})}{\prod_{k=1}^{10} \dim j_k}$$

is a bounded function in $\mathbb{R}_+^{10}$. Therefore, there exists $C > 0$ such that

$$\frac{|W(j, \vec{n})|}{\prod_{k=1}^{10} \dim j_k} \leqslant C. \tag{3.28}$$

This bound can be rewritten as

$$|W(j, \vec{n})| \leqslant C \prod_{k=1}^{10} \dim j_k, \tag{3.29}$$

which is convenient for investigating the absolute convergence of the state sum.

### 3.3.2 The finiteness bounds

We showed in the previous section that the vertex amplitude divided by the product of the dimensions of the vertex spins is a bounded function of spins. This result suggests to introduce a rescaled vertex amplitude $W_p$ as

$$W_p(j_f, \vec{n}_{ef}) = \frac{W(j_f, \vec{n}_{ef})}{\prod_{f=1}^{10}(\dim j_f)^p}, \tag{3.30}$$

where $p \geqslant 0$, in order to improve the convergence of the state sum.

Given a triangulation $T$ of a compact four-manifold $M$, we will consider the following state sum

$$Z_p = \sum_{j_f} \int \prod_{e,f} d^2\vec{n}_{ef} \prod_{f \in T^*} \dim j_f \prod_{v \in T^*} W_p(j_{f(v)}, \vec{n}_{e(v)f(v)}). \tag{3.31}$$

It is sufficient to consider $T$ without a boundary, since if $Z(T)$ is finite, then $Z(\gamma, T)$ will be finite due to gluing properties, where $\gamma$ is the boundary spin network.

The convergence of $Z_p$ will be determined by the large-spin asymptotics of the vertex amplitude $W$ and the values of $p$. Since the asymptotics of $W$ is not known completely, we will use the estimate (3.29) in order to find the values of $p$ which make the state sum $Z_p$ convergent. Since

$$|Z_p| \leqslant \sum_{j_f} \int \prod_{e,f} d^2\vec{n}_{ef} \prod_{f \in T^*} \dim j_f \prod_{v \in T^*} \frac{|W(j_{f(v)}, \vec{n}_{e(v)f(v)})|}{\prod_{f \in v}(\dim j_{f(v)})^p}, \tag{3.32}$$

and by using (3.29) we obtain

$$|Z_p| \leqslant C^V \sum_{j_f} \int \prod_{e,f} d^2\vec{n}_{ef} \prod_{f \in T^*} \dim j_f \prod_{v \in T^*} \frac{1}{\prod_{f \in v}(\dim j_{f(v)})^{p-1}},$$

where $V$ is the total number of vertices in the triangulation $T$. At this point the integrand does not depend anymore on $\vec{n}_{ef}$, so the appropriate integration over $4E$ 2-spheres can be performed. Here $E$ is the total number of edges in $\sigma$, and it is multiplied by 4 since every edge is a boundary for exactly four faces. After the integration we obtain

$$|Z_p| \leqslant C^V (4\pi)^{4E} \sum_{j_f} \prod_{f \in T^*} \dim j_f \prod_{v \in T^*} \frac{1}{\prod_{f \in v}(\dim j_{f(v)})^{p-1}}. \tag{3.33}$$

The sum over the spins in (3.33) can be rewritten as a product of single-spin sums. Let $N_f$ be the number of vertices bounding a given face $f$. Each vertex contributes with a factor $(\dim j_f)^{-p+1}$, so the total contribution for each face $f$ is $(\dim j_f)^{1-(p-1)N_f}$. Thus we can rewrite (3.33) as

$$|Z_p| \leqslant C^V (4\pi)^{4E} \prod_{f \in T^*} \sum_{j_f \in \frac{N_0}{2}} (\dim j_f)^{1-(p-1)N_f} . \qquad (3.34)$$

The sum in (3.34) will be convergent if

$$1 - (p-1)N_f < -1 ,$$

or

$$p - 1 > \frac{2}{N_f} \qquad (3.35)$$

for every $N_f$. Since $N_f \geqslant 2$ for every face $f$, a sufficient condition for $p$ is

$$p > 2 . \qquad (3.36)$$

Therefore $Z_p$ is absolutely convergent for $p > 2$, which means that it is convergent for $p > 2$. As far as the convergence of $Z_p$ for $p \leqslant 2$ cases is concerned, one has to calculate the ranks of the Hessians $H_m$ and use the following inequalities

$$|\det(-\tilde{H}_m)| \geqslant C_m \left( \prod_{k=1}^{m} \dim j_k \right)^{r/m} , \qquad (3.37)$$

when possible. We expect that the inequalities (3.37) will hold for all $m$, since $\det(-\tilde{H}_m)$ is a homogeneous polynomial of the spins of the degree $r$ and $Re\,(-\tilde{H}_m)$ is a positive definite matrix. Then

$$|W(j, \vec{n})| \leqslant C_q \left( \prod_{k=1}^{10} \dim j_k \right)^{1-q} , \qquad (3.38)$$

for any $j$, where $q = \min\{r/2m \,|\, m = 3, \ldots, 10\}$. Since $q > 0$, the new bound (3.38) will be an improvement of the bound (3.29) and consequently $Z_p$ will be absolutely convergent for

$$p > 2 - q . \qquad (3.39)$$

Given that $r = 44$ for $m = 10$, this implies that $q \geqslant 1/18$ ($r = 1$ and $m = 9$ case) and therefore $p > 35/18$.

Finally, in order to investigate the absolute or a simple convergence for $p \leqslant 35/18$ one has to devise some other methods.

### 3.4 Path integral for $BFCG$ theory

Given the $BFCG$ form of the EC action, one can now proceed to quantize the theory by using the same approach as in the case of spin-foam models, see [113]. This approach requires first a construction of the state-sum model for the topological theory given by the unconstrained $BFCG$ theory. Then the constraint $B_{ab} = \varepsilon_{abcd}\, e^c \wedge e^d$ has to be imposed on the topological state sum.

In the topological case one starts from the path-integral

$$
\begin{aligned}
Z &= \int \mathcal{D}A\, \mathcal{D}\beta\, \mathcal{D}B\, \mathcal{D}C \, \exp\left( i \int_M (\langle B \wedge \mathcal{F} \rangle + \langle C \wedge \mathcal{G} \rangle) \right) \\
&= \int \mathcal{D}A\, \mathcal{D}\beta\, \delta(\mathcal{F})\, \delta(\mathcal{G}),
\end{aligned}
\tag{3.40}
$$

see [53]. Let $T$ be a regular triangulation of $M$ and $T^*$ the corresponding dual triangulation. Then

$$
Z = \int \prod_l dg_l \int \prod_f dh_f \prod_f \delta(g_f) \prod_p \delta(h_p),
\tag{3.41}
$$

where $l, f$ and $p$ denote the 1-, 2- and 3-cells of $T^*$, respectively, and one has

$$
g_l = \exp\left( \int_l A \right), \quad h_f = \exp\left( \int_f \beta \right).
$$

The group elements $g_l \in G$ and $h_f \in H$ represent the corresponding 1- and 2-holonomies of $A$ and $\beta$, respectively. The group element $g_f = \prod_{l \in \partial f} g_l$ is the holonomy along the boundary of $f$. When the area of $f$ is small, one has

$$
g_f \approx \exp\left( \int_f \mathcal{F} \right).
$$

The group element $h_p$ is the 2-holonomy along the closed surface $\partial p$, and

$$
h_p = \prod_{f \in \partial p} \tilde{h}_f,
$$

where some of the $\tilde{h}_f$ are given by $g_l \triangleright h_f$, where $l \in p$ and $l \notin \partial f$, while the other $\tilde{h}_f$ are equal to $h_f$, see [53]. When the volume of $p$ is small, one has

$$
h_p \approx \exp\left( \int_p \mathcal{G} \right).
$$

In the case of the Poincaré 2-group the integral (3.41) can be written as

$$Z = \int \prod_l dg_l \int \prod_f d^4 \vec{x}_f \prod_f \delta(g_f) \prod_p \delta(\vec{x}_p) , \qquad (3.42)$$

where $\vec{x}_p = \vec{x}_f + \cdots + g_l \vec{x}_{f'}$ and $f, \ldots, f' \in \partial p$. The Lorentz group delta function can be expanded by using the Plancherel theorem

$$\delta(g_f) = \sum_{\Lambda_f} d\mu(\Lambda_f) \chi(g_f, \Lambda_f) ,$$

where $\Lambda = (j, \rho)$ are the unitary irreducible representations, $\chi$ is the character and $d\mu$ is the appropriate integration measure, see [115]. The notation $\delta(g)$ means that the corresponding distribution is concentrated at the identity element of the Lorentz group. The $\delta(\vec{x}_p)$ is the four-dimensional Dirac delta function and

$$\delta(\vec{x}_p) = \frac{1}{(2\pi)^4} \int_{\mathbb{R}^4} d^4 \vec{L}_p \exp\left( i\vec{x}_p \cdot \vec{L}_p \right) .$$

We will take that the components of $\vec{L}_p$ have dimensions of length, so that the components of $\vec{x}_p$ will have dimensions of length divided by Planck's constant $\hbar$ in order for $\vec{x}_p \cdot \vec{L}_p$ to be dimensionless.

For the sake of simplicity, let us consider the Euclidean case, so that the Poincaré 2-group is replaced by the Euclidean 2-group. The Lorentz group is then replaced by the $SO(4)$ group and $\Lambda = (j^+, j^-)$ is a pair of $SU(2)$ spins, so that

$$Z = \sum_{\Lambda_f} \int \prod_p d^4 \vec{L}_p \int \prod_l dg_l \prod_f d^4 \vec{x}_f \, \dim \Lambda_f \, \chi(\Lambda_f, g_f) \prod_p e^{i\vec{x}_p \cdot \vec{L}_p} .$$

$$(3.43)$$

After integrating $\vec{x}_f$, we obtain

$$Z = \sum_{\Lambda_f} \int \prod_p d^4 \vec{L}_p \int \prod_l dg_l \prod_f \dim \Lambda_f \, \chi(\Lambda_f, g_f)$$
$$\prod_f \delta\left( g_{l_1(p_1,f)} \vec{L}_{p_1,f} + g_{l_2(p_2,f)} \vec{L}_{p_2,f} + g_{l_3(p_3,f)} \vec{L}_{p_3,f} \right) , \qquad (3.44)$$

where $p_1, p_2$ and $p_3$ are the three polyhedra which share the face $f$, and $l_1, l_2$ and $l_3$ are the corresponding dual edges satisfying $l_k \in p_k$ and $l_k \notin f$. It is instructive to rewrite (3.44) by using the simplices of $T(M)$

$$Z = \sum_{\Lambda_\Delta} \int \prod_\varepsilon d^4 \vec{L}_\varepsilon \int \prod_\tau dg_\tau \prod_\Delta \dim \Lambda_\Delta \, \chi(\Lambda_\Delta, g_\Delta)$$
$$\prod_\Delta \delta\left( g_{\tau_1(\varepsilon_1,\Delta)} \vec{L}_{\varepsilon_1,\Delta} + g_{\tau_2(\varepsilon_2,f)} \vec{L}_{\varepsilon_2,\Delta} + g_{\tau_3(\varepsilon_3,\Delta)} \vec{L}_{\varepsilon_3,\Delta} \right) , \qquad (3.45)$$

where $\varepsilon_k$ are the edges of $\Delta$ and $\varepsilon_k \in \tau_k$ but $\Delta \notin \tau_k$. The delta function

$$\delta(g_1 \vec{L}_1 + g_2 \vec{L}_2 + g_3 \vec{L}_3)$$

in (3.45) restricts the integration over the $\vec{L}$'s whose lengths satisfy the triangle inequalities. This implies that $L_\varepsilon \equiv |\vec{L}_\varepsilon|$ can be interpreted as the length of an edge $\varepsilon$. Note that (3.45) can be rewritten as

$$Z = \int \prod_\varepsilon L_\varepsilon^3 dL_\varepsilon \sum_{\Lambda_\Delta, I_\tau} W(L, \Lambda, I), \qquad (3.46)$$

where $I_\tau$ is the intertwiner for the four $\Lambda$ of a tetrahedron and

$$\sum_I W(L, \Lambda, I) = \int \prod_\varepsilon d\Omega_\varepsilon \int \prod_\tau dg_\tau \prod_\Delta \dim \Lambda_\Delta \, \chi(\Lambda_\Delta, g_\Delta)$$

$$\prod_\Delta \delta \left( g_{\tau_1(\varepsilon_1, \Delta)} \vec{L}_{\varepsilon_1, \Delta} + g_{\tau_2(\varepsilon_2, f)} \vec{L}_{\varepsilon_2, \Delta} + g_{\tau_3(\varepsilon_3, \Delta)} \vec{L}_{\varepsilon_3, \Delta} \right). \quad (3.47)$$

The $d\Omega_\varepsilon$ denotes a 3-sphere volume measure.

The state sums/integrals in (3.46) will be almost certainly divergent, but what is important is to find the structure of the dual 3-complex amplitude $W$. The relation (3.47) seems to imply that $W$ will be a function of the $SO(4)$ irreps, but this may not happen because the integral in (3.47) may not be well-defined and a regularization may introduce the irreps for $SO(3)$ and $SO(2)$ subgroups. That this can happen is suggested by the representation theory of the Poincaré/Euclidean 2-group on 2-Hilbert spaces, see [7, 37]. Namely, in the Euclidean case the irreps are labeled by positive numbers, which can be identified with the edge lengths $L_\varepsilon$. In the Poincaré 2-group case there is also a class of positive-length irreps, and in both cases the corresponding triangle intertwiners are the $SU(2)$ spins when $L_\varepsilon$ form a zero-area triangle. When $L_\varepsilon$ form a non-zero area triangle, then the intertwiners are given by the $U(1)$ spins. The results of [10, 11] suggest that the topological amplitude is given by

$$W(L, m, I) = \prod_f A_f(L) \prod_v \frac{\cos S_v(L, m)}{V_v(L)}, \qquad (3.48)$$

where

$$S_v(L, m) = \sum_{f \ni v} m_f \theta_f(L). \qquad (3.49)$$

The angle $\theta_f(L)$ is the interior dihedral angle for a face $f$ which contains the vertex $v$, $V_v(L)$ is the 4-volume of the four-simplex dual to $v$, $A_f(L)$ is the area of the triangle dual to $f$, $m_f \in \mathbb{Z}$ are the $U(1)$ spins and the

2-intertwiners $I$ are trivial. Note that taking the sum of (3.49) over all vertices gives

$$\sum_{v \in T^*} S_v(L, m) = \sum_{v \in T^*} \sum_{f \ni v} m_f \theta_{f,v}(L)$$

$$= \sum_{f \in T^*} m_f \left( \sum_{v \in f} \theta_{f,v}(L) \right)$$

$$= \sum_{f \in T^*} m_f \delta_f(L)$$

$$= S_R(L, m), \tag{3.50}$$

where $\delta_f(L) = \sum_{v \in f} \theta_{f,v}(L)$ is the deficit angle for the triangle dual to the face $f$, and $S_R$ is the total Regge-like action over the triangulation $T$. It differs from the usual Regge action by the presence of the integer $m_f$ in place of the area of the triangle, $A_f(L)$ — the former is an integer independent of edge lengths, while the latter is a function of edge lengths given by the Heron formula.

Given the topological amplitude (3.48), one can try to implement the constraint $B^{ab} = \varepsilon^{abcd} e_c \wedge e_d$ in order to obtain the state sum for GR, similarly to what was done in the case of spin foam models. Note that $S_v$, given by equation (3.49), has the form of the area-Regge action for a 4-simplex. If we put

$$|m_f| l_P^2 = A_f(L), \tag{3.51}$$

where $l_P$ is the Planck length, then the area-Regge action $S_v(L, m)$ will become the Regge action $S_{vR}(L)$ for a four-simplex. Since the topological vertex amplitude (3.48) is proportional to $\cos S_v$, which is a sum of $e^{iS_v}$ and $e^{-iS_v}$, then in the state sum amplitude will appear a term proportional to

$$\prod_v e^{iS_v(L)} = \exp\left( i \sum_v S_v(L) \right) = \exp\left( \frac{i}{l_P^2} \sum_f A_f(L) \delta_f(L) \right),$$

where $\delta_f$ is the deficit angle and $S_R = \frac{1}{l_P^2} \sum_f A_f(L) \delta_f(L)$ is the Regge action for a manifold triangulation. The appearance of the term in the amplitude proportional to $e^{iS_R}$ is a good sign that the constrained state sum can be further modified such that it corresponds to the path integral for GR. Therefore we expect that the quantum GR state sum will have a form

$$Z_{GR} = \int \prod_\varepsilon \mu(L_\varepsilon)\, dL_\varepsilon \sum_m \prod_\Delta \delta(|m_\Delta| l_P^2 - A_\Delta(L))\, W_{GR}(L, m). \quad (3.52)$$

The amplitude $W_{GR}$ and the measure $\mu$ have to be chosen such that $Z_{GR}$ is finite and that the corresponding effective action gives GR in the classical limit. The effective action approach to the semiclassical limit of spin-foam models [94] suggests that

$$W_{GR}(L, m) = \prod_f A_f(L) \prod_v \frac{e^{iS_v(L,m)}}{V_v(L)}, \quad (3.53)$$

i.e., the $\cos S_v$ factor from (3.48) has to be replaced by $e^{iS_v}$ so that the effective action will have the correct classical limit (see also [92, 124]).

Coupling of matter in the model defined by the state sum (3.52) will be easier than in the EPRL/FK model case [124], since the edge lengths $L_\varepsilon$ are explicitly present. One can then use

$$W_{\text{matt}}(L, m, \varphi) \propto \exp\left( iS_R^{(\text{matt})}(L, \varphi)/l_P^2 \right),$$

for the matter amplitudes, where $S_R^{(\text{matt})}$ is the Regge discretized action of a matter field $\varphi$ coupled to gravity. The expressions for $V_\tau(L)$ and $V_\sigma(L)$, which appear in $S_R^{(\text{matt})}$, can be easily obtained, in contrast to the EPRL/FK model case, where the expression for $V_\sigma$ is difficult to write explicitly in terms of the spin foam variables.

As far as the boundary states are concerned, one will have a wavefunction $\Psi(L_\varepsilon, m_\Delta)$ on the boundary $\partial M = \Sigma$, where $\varepsilon, \Delta \in T(\Sigma)$ and $T(\Sigma)$ is the triangulation of the three-dimensional manifold $\Sigma$ induced by $T(M)$. By passing to the dual complex $T^*(\Sigma)$, the wavefunction can be written as $\Psi(L_f, m_l)$, i.e. a function of a colored 2-complex. This reflects the fact that a boundary of a colored 3-complex is a colored 2-complex.

## 3.5 Spin-cube models

As explained in section 3.2, the spin foam models are discrete path-integral formulations of gauge theories and quantum gravity, also see [6, 113]. The path integral for a spin foam model is defined as a state sum for a colored dual 2-complex of the spacetime manifold triangulation and the colors are chosen to be the objects and the morphisms of a representation category of the relevant symmetry group. In the case of GR this group is the Lorentz group. A natural categorical generalization of a spin-foam model would be a state-sum model based on a colored 3-complex, where the colors are the objects, morphisms and 2-morphisms of a 2-category representation of the relevant 2-group, see [11, 95, 98, 123]. We will refer to these models as spin-cube models, and in the case of GR, the relevant 2-groups are the Poincaré 2-group [95] and the teleparallel 2-group [9].

If one labels the 3-cells, 2-cells and 1-cells of a given 3-complex with the objects, morphisms and 2-morphisms of a given 2-category, this is equivalent to labeling the edges, triangles and tetrahedrons of a spacetime triangulation. Hence the spin-cube models give a possibility of introducing the edge lengths as degrees of freedom, beside the triangle spins and the tetrahedron intertwiners, which are the spin foam variables. In the case of the Poincaré 2-group there is a representation 2-category such that the objects (representations) are labeled by positive numbers. These representations satisfy the triangle inequalities when composed and the corresponding intertwiners are $U(1)$ spins for non-zero area triangles [7, 37].

The reason why one would like to introduce the edge lengths as additional degrees of freedom, is that in this way one can solve the problems of spin-foam models related with the fact that an arbitrary spin-foam configuration does not correspond to a metric geometry. Namely, the spins of triangles in a spin-foam model correspond to the areas of triangles, and as we saw in section 2.3, an arbitrary assignment of triangle areas does not give a well-defined metric geometry, see also [16, 51, 129], unless the edge-length constraints are imposed, also see [75]. The edge-length constraints can be replaced by the Dittrich-Speziale (DS) constraints, see section 2.3.1, and in the current formulations of spin-foam models [44, 49], the imposition of the DS constraints is made in an indirect manner, see section 3.3, so that there are no guarantees that the correct semiclassical limit will be obtained. Actually, as we will see in section 4.3, one does not obtain the correct semiclassical limit.

In addition, it is difficult to couple fermionic matter to spin-foam models, since the fermions couple to the tetrads, and these are not well-defined in an arbitrary spin-foam configuration. Also, when the effective action is computed in the semiclassical approximation, see section 4.3, the classical limit is the area-Regge action, also see [94, 96]. Hence the classical limit of spin-foam models for smooth spacetimes cannot be automatically identified with the Einstein-Hilbert action. The explicit presence of the edge-length variables in spin-cube models solves automatically the problem of coupling of fermionic matter, since the tetrads can be easily defined when the edge-lengths are given, see section 2.2, while the effective action for a spin-cube model can naturally have the usual Regge action as its classical limit, see section 4.4.

### 3.5.1 *Poincaré 2-group state-sum models*

A 2-group is a categorification of a group, since a group is an invertible category with one object, while a 2-group is an invertible 2-category with one object, see [8]. Any strict[3] 2-group is equivalent to a crossed module, and the latter is simply a pair of groups $G$ and $H$ such that there is a map $\partial : H \to G$ which is a homomorphism and a map $\triangleright : G \times H \to H$, which is a group action, such that

$$\partial(g \triangleright h) = g(\partial h)g^{-1}, \quad (\partial h) \triangleright h' = hh'h^{-1},$$

where $g \in G$ and $h, h' \in H$. See Appendix A for a formal definition. A typical example is the $n$-dimensional Euclidean 2-group, where $G = SO(n)$ and $H = \mathbb{R}^n$. The $\partial$ map is trivial while the $\triangleright$ map is the usual action of a rotation on a vector. Another example is the Poincaré 2-group — the semi-direct product $G \times_s H$ corresponds to the group of 2-morphisms in a 2-group, so that the usual Poincaré group is only a part of the Poincaré 2-group where $G = SO(3, 1)$ and $H = \mathbb{R}^4$.

The reason why the Poincaré 2-group is relevant for GR is that GR can be represented as a gauge theory for the Poincaré 2-group [95, 98, 123]. More precisely, the Einstein equations can be derived from an action which describes a constrained $2BF$ theory for the Poincaré 2-group

$$S = \int_M \left[ B^{ab} \wedge R_{ab} + e^a \wedge \nabla \beta_a - \lambda^{ab}(B_{ab} - \varepsilon_{abcd} \, e^c \wedge e^d) \right], \qquad (3.54)$$

where $R_{ab}$ is the curvature 2-form for the Lorentz group connection $\omega_{ab}$ and $\beta_a$ is a 2-form which together with $\omega_{ab}$ forms a 2-connection $(\omega_{ab}, \beta_a)$ for the Poincaré 2-group. The 2-forms $B_{ab}$ and the 1-forms $e_a$, which can be identified with the tetrads, enforce the vanishing of the 2-curvature

$$(R_{ab}, \nabla \beta_a) \equiv (\, d\omega_{ab} + \omega_{ac} \wedge \omega^c{}_b \, , \, d\beta_a + \omega_{ab} \wedge \beta^b \,),$$

in the topological case, when $\lambda_{ab} = 0$. The constraint

$$B_{ab} = \varepsilon_{abcd} \, e^c \wedge e^d, \qquad (3.55)$$

---

[3]In a strict category, all the defining relations between the elements of the category are the identity maps.

transforms the topological gravity theory

$$S_{top} = \int_M \left( B^{ab} \wedge R_{ab} + e^a \wedge \nabla \beta_a \right),$$

into GR and it is the same constraint which is used in the case of spin-foam models. However, in the Poincaré 2-group case the GR constraint can be written in a simpler way since the tetrads appear explicitly in the theory.

A quantum gravity theory can be constructed by using the path integral based on the action (3.54), see [95]. This theory takes a form of a state-sum model for a colored dual 3-complex of a triangulation of the spacetime manifold. The set of colors consists of positive numbers for the edges, which satisfy the triangle inequalities, while the colors for the triangles and the tetrahedrons can be the irreps and the corresponding intertwiners for the Lorentz group or its $SO(3)$ and $SO(2)$ subgroups.

This result agrees with the categorical structure of a state sum for a 2-group, since the labels for the edges can be interpreted as the labels for 2-group representations, while the labels for the triangles can be interpreted as the corresponding intertwiners. The labels for the tetrahedrons can be interpreted as the 2-intertwiners, and they arise because a 2-group representation category is a 2-category, and hence the 2-intertwiners correspond to 2-morphisms.

In the Poincaré/Euclidean 2-group case there is a 2-Hilbert space representation 2-category, see [7,37], such that the object (representation) labels are positive numbers. The corresponding triangle intertwiners are $SO(2)$ or $U(1)$ irreps if the triangles have non-zero areas. The 2-intertwiner labels for the tetrahedra are trivial, so that one can construct a state sum as

$$Z = \int_{\tilde{\mathbb{R}}_+^E} \prod_{\epsilon=1}^{E} \mu(L_\epsilon) \, dL_\epsilon \sum_{m \in \mathbb{Z}^F} \prod_{\Delta=1}^{F} W_\Delta(L, m) \prod_{\sigma=1}^{V} W_\sigma(L, m), \qquad (3.56)$$

where $\epsilon$ are the edges of a triangulation $T(M)$ of the 4-manifold $M$, $\Delta$ are the triangles of $T(M)$ and $\sigma$ are the 4-simplices of $T(M)$. $E$ is the number of edges, $F$ is the number of triangles, $V$ is the number of 4-simplices and $\tilde{\mathbb{R}}_+^E$ is the subset of $\mathbb{R}_+^E$ whose elements satisfy the triangle inequalities associated with the triangulation $T(M)$.

The weights $\mu_\epsilon$, $W_\Delta$ and $W_\sigma$ should be chosen such that the state sum $Z$ resembles a discretized path integral for GR. More precisely, a choice of the weights should be such that it implements the GR constraint (3.55) and that the corresponding state-sum model defines a quantum gravity theory whose classical limit is the Regge action

$$S_R = \sum_{\Delta=1}^{F} A_\Delta(L)\,\theta_\Delta(L)\,, \tag{3.57}$$

where $A_\Delta$ is the area of a triangle $\Delta$ and $\theta_\Delta$ is the deficit angle. We will refer to (3.57) as the length-Regge action in order to distinguish it from the area-Regge action

$$S_{AR} = \sum_{\Delta=1}^{F} A_\Delta\,\tilde{\theta}_\Delta(A)\,, \tag{3.58}$$

which can be naturally associated to a spin-foam model.

### 3.5.2 *Poincaré 2-group state sum for GR*

The GR constraint (3.55) can take the following form in the discrete setting

$$\gamma m_\Delta = A_\Delta(L)/l_P^2\,, \tag{3.59}$$

where $m_\Delta \in \mathbb{N}$ is an $SO(2)$ spin of a triangle $\Delta$, $A_\Delta(L)$ is the area of a triangle with edge lengths $L_1, L_2$ and $L_3$ and $\gamma$ is a constant, which is analogous to the Barbero-Immirzi constant which appears in the case of spin-foam models. In order to have simpler formulas, we are going to take $\gamma = 1$. The function $A(L)$ is given by Heron's formula

$$A(L) = \sqrt{s(s - L_1)(s - L_2)(s - L_3)}\,, \tag{3.60}$$

where $2s = L_1 + L_2 + L_3$ is the triangle perimeter.

In order to get physical lengths and areas, one has to make the rescaling $L \to L/l_0$ in (3.59), where $l_0$ is a unit of length. It is natural to choose $l_0$ to be the Planck length $l_P$. Note that choosing $l_0$ to be a multiple of $l_P$ is equivalent to choosing $\gamma \neq 1$.

The constraints (3.59) can be implemented in the state sum (3.56) by choosing the triangle weights as

$$W_\Delta = \delta\left(m_\Delta - \frac{A_\Delta(L)}{l_P^2}\right)\,. \tag{3.61}$$

In order to ensure that the Regge action will be the classical limit of the model, we will choose

$$W_\sigma = \exp\left(i \sum_{\Delta \in \sigma} m_\Delta\,\theta_\Delta^{(\sigma)}(L)\right)\,, \tag{3.62}$$

where $\theta_\Delta^{(\sigma)}(L)$ is the interior dihedral angle [95]. The reason for this choice is simple to understand, since

$$\prod_{\sigma=1}^{V} \exp\left( i \sum_{\Delta \in \sigma} m_\Delta \, \theta_\Delta^{(\sigma)}(L) \right) = \prod_{\sigma=1}^{V} \exp\left( i \sum_{\Delta \in \sigma} A_\Delta(L) \, \theta_\Delta^{(\sigma)}(L)/l_P^2 \right),$$

due to the constraint $m_\Delta = A_\Delta(L)$, so that

$$\prod_{\sigma=1}^{V} \exp\left( i \sum_{\Delta \in \sigma} A_\Delta(L) \, \theta_\Delta^{(\sigma)}(L)/l_P^2 \right) = e^{i S_R(L)/l_P^2}.$$

Hence the constraints (3.59) can reduce the spin-cube state sum to a path integral for the Regge model. However, there are certain caveats in this simple reasoning, which we will demonstrate by a more precise analysis. Let us start from the state sum with the weights (3.61) and (3.62)

$$Z = \sum_{m \in \mathbb{N}^F} \int_{\tilde{\mathbb{R}}_+^E} \prod_{\epsilon=1}^{E} \mu(L_\epsilon)\, dL_\epsilon \prod_{\Delta=1}^{F} \delta(m_\Delta - A_\Delta(L)/l_P^2)$$
$$\prod_{\sigma=1}^{V} \exp\left( i \sum_{\Delta \in \sigma} m_\Delta \, \theta_\Delta^{(\sigma)}(L) \right). \tag{3.63}$$

The form of (3.63) suggests to integrate first the lengths, which will transform (3.63) into a sum over the spins subject to the constraints

$$m_f - \frac{A_f(L)}{l_P^2} = 0, \quad f = 1, 2, \ldots, F. \tag{3.64}$$

In order to solve these constraints, note that in a four-manifold triangulation we have

$$F \geqslant \frac{4}{3} E,$$

since $F$ triangles have $3F$ edges, and each edge is shared by at least 4 triangles, so that $3F \geqslant 4E$. Consequently

$$F > E,$$

so that we can solve the first $E$ constraints of (3.64) as

$$L_\epsilon = l_\epsilon(m_1, \ldots, m_E), \tag{3.65}$$

where $\epsilon = 1, 2, \ldots, E$, while the remaining $F - E$ constraints become the Diophantine equations

$$m_k = \varphi_k(m_1, \ldots, m_E), \quad E + 1 \leqslant k \leqslant F, \tag{3.66}$$

where $\varphi_k(m) = A_k(l(m))$. Hence $m \in D_F \subset \mathbb{N}^F$. However, it is difficult to determine the structure of $D_F$ and a priori it may even be an empty set, although see also section 3.5.4.

This problem can be solved by relaxing the constraints (3.66) as

$$m_k = [\varphi_k(m_1, \ldots, m_E)], \quad E+1 \leqslant k \leqslant F, \tag{3.67}$$

where $[x]$ is the integer part of a real number $x$. In this case the constraints are given by

$$m_e = A_e(L), \quad 1 \leqslant e \leqslant E,$$

$$m_k = [A_k(L)], \quad E+1 \leqslant k \leqslant F, \tag{3.68}$$

and the solution is $L_\epsilon = l_\epsilon(m')$ where $m' \in \mathbb{N}^E$ and $m'' = [\varphi(m')] \in \mathbb{N}^{F-E}$. Since the functions $l_\epsilon(m')$ have to be real, this means that $m' \in D_E \subset \mathbb{N}^E$, which is related to the fact that $L_\epsilon$ have to satisfy the triangle inequalities.

Let us now introduce the new weights in the spin-cube state sum, so that we start from (3.56) with

$$\prod_{\Delta=1}^{F} W_\Delta(L, m) = \prod_{f=1}^{E} \delta(m_f \quad A_f(L)/l_P^2) \prod_{f=E+1}^{F} \delta(m_f l_P^2 - [A_f(L)/l_P^2]) \tag{3.69}$$

and $W_\sigma$ is given by (3.62). By integrating the $L$ variables we obtain the following state-sum model

$$Z = \sum_{m \in D_E} \prod_{\epsilon=1}^{E} \mu_\epsilon(l(m)) \, J(m_1, \ldots, m_E)$$

$$\exp\left(i \sum_{f=1}^{E} m_f \theta_f(m) + i \sum_{f=E+1}^{F} [\varphi_f(m)] \theta_f(m)\right), \tag{3.70}$$

where

$$J(m_1, \ldots, m_E) = \left| \frac{\partial(L_1, \ldots, L_E)}{\partial(m_1, \ldots, m_E)} \right|$$

is the Jacobian for $L_\epsilon = l_\epsilon(m)$. Note that this is a state-sum model with a non-local weight

$$W_E(m) = \prod_{\epsilon=1}^{E} \mu_\epsilon(l(m)) \, J(m_1, \ldots, m_E) \tag{3.71}$$

and the state sum has a form of a path integral for an area-Regge model

$$Z = \sum_{m \in D_E} W_E(m) \exp\left(i S_{AR}^*(m)\right),$$

where

$$S_{AR}^*(m) = \sum_{f=1}^{E} m_f \theta_f(m) + \sum_{f=E+1}^{F} [\varphi_f(m)] \theta_f(m).$$

This is an area-Regge action, with integer areas, where the edge-length constraints are imposed via (3.67). The finiteness and the effective action for the state sum model (3.70) can be studied by using the techniques of [94, 96, 97]. We will not do this here, since the analysis gets complicated due to the presence of the non-local weight (3.71).

Note that one can define a new model, by choosing $\mu(L_e) = 1$, $W_\sigma$ as in (3.62) and a non-local weight for the triangles in the spin-cube state sum

$$\tilde{W}(L, m) = J^{-1}(m_1, \ldots, m_E) \prod_{\Delta=1}^{F} W_\Delta(L, m) \prod_{\Delta=1}^{E} m_\Delta^{-p},$$

where $W_\Delta$ are given by (3.69). This choice of the weights gives a state-sum model with local weights for the triangles

$$\tilde{Z} = \sum_{m \in D_E} \prod_{f=1}^{E} m_f^{-p} \exp\left(iS_{AR}^*(m)/l_P^2\right). \tag{3.72}$$

The semiclassical effective action for the area-Regge spin-foam model (3.72) can be easily calculated by using the results of [94, 96], and also see chapter 4. We obtain for $m \to \infty^E$

$$\Gamma(m) = S_{AR}^*(m)/l_P^2 + p \sum_{f=1}^{E} \ln m_f + \frac{1}{2} \operatorname{Tr}\left(\log(S_{AR}^*)''(m)\right) + O(m^{-2}), \tag{3.73}$$

where $(S_{AR}^*)''(m)$ is the Hessian matrix for the function $(S_{AR}^*)(m)$. Since

$$S_{AR}^*(m) = O(m), \qquad p \sum_{f=1}^{E} \ln m_f = O(\ln m),$$

$$\operatorname{Tr}\left(\log(S_{AR}^*)''(m)\right) = O(m^{-1}), \tag{3.74}$$

where the notation $f(m) = O(m^r)$ means that

$$f(\lambda m_1, \ldots, \lambda m_E) \approx \lambda^r g(m, \lambda)$$

and $f(m) = O(\ln m)$ means

$$f(\lambda m_1, \ldots, \lambda m_E) \approx (\ln \lambda) g(m, \lambda)$$

for $\lambda \to \infty$ and $g(m, \lambda)$ is a bounded function of $\lambda$. From (3.74) it follows that the classical limit of the effective action (3.73) will be the area-Regge action $S_{AR}^*(m)$. However, the action $S_{AR}^*(m)$ is dynamically equivalent to the length-Regge action $S_R(L)$ due to the constraints (3.67).

As far as the convergence of the state sum (3.72) is concerned, it is easy to see that it is absolutely convergent for $p > 1$, while the convergence for $p \leqslant 1$ case is a more complicated issue and will not be analyzed here.

### 3.5.3 Edge-length state-sum models

The state-sum model (3.70) appeared because we first integrated the edge-lengths in the spin-cube state sum. This was a natural way to proceed, because of the delta-function weights (3.61) and the fact that the spins $m$ are integers. A natural question to ask is it possible to implement the constraints such that the edge lengths remain as the independent variables.

A clue comes from the relaxed constraints (3.68), so let us instead consider the following set of constraints

$$m_f = [A_f(L)/l_P^2], \quad f = 1, 2, \ldots, F. \tag{3.75}$$

These constraints have solutions for any $L \in \tilde{\mathbb{R}}_+^E$, and if we take

$$W_f(L, m) = \delta\left(m_f - [A_f(L)/l_P^2]\right),$$

with $W_\sigma$ given by (3.62), then the summation over the spins $m$ in (3.56) gives

$$Z = \int_{\tilde{\mathbb{R}}_+^E} \prod_{\epsilon=1}^{E} \mu_\epsilon(L) \, dL_\epsilon \, \exp\left(i\tilde{S}_R(L)/l_P^2\right), \tag{3.76}$$

where

$$\tilde{S}_R = \sum_{\Delta=1}^{F} [A_\Delta(L)] \theta_\Delta(L).$$

Hence the constraints (3.75) reduce the state sum to a path integral for a continuous-length integer-area Regge model. The measure $\mu$ can be chosen such that $Z$ is finite. For example

$$\mu(L_\epsilon) = \left(1 + \frac{L_\epsilon}{l_0}\right)^{-p}, \tag{3.77}$$

where $p > 0$ and $l_0 > 0$. This measure will give an absolutely convergent partition function for $p > 1$, since

$$|Z| \leqslant \int_{\tilde{\mathbb{R}}_+^E} \prod_{\epsilon=1}^{E} \left(1 + \frac{L_\epsilon}{l_0}\right)^{-p} dL_\epsilon < \int_{\mathbb{R}_+^E} \prod_{\epsilon=1}^{E} \left(1 + \frac{L_\epsilon}{l_0}\right)^{-p} dL_\epsilon,$$

so that

$$|Z| < \left(\int_0^{+\infty} \frac{dL}{\left(1 + \frac{L}{l_0}\right)^p}\right)^E. \tag{3.78}$$

The integral in (3.78) is convergent for $p > 1$. More generally, $\mu$ can be chosen such that $\mu(0)$ is finite and $\mu(L) = O(L^{-p})$ where $p \in \mathbb{R}$. However, the convergence of the state sum for $p \leqslant 1$ case is a more complicated problem and we will not attempt to resolve it here.

### 3.5.4　*Imposing the simplicity constraint*

As explained in chapter 2, given a classical action, one constructs the state-sum model in the following way. First, one rewrites the classical action into a topological sector and the simplicity constraint. Second, one constructs the state sum model for the topological sector of the theory, so that the resulting state sum is a triangulation independent topological invariant. Third, one modifies the amplitudes and imposes the simplicity constraint in a certain way, in order to pass from the topological state-sum model to the non-topological state-sum model which represents the quantization of the full classical action.

The crucial difference between the topological and non-topological state-sum models lies in the implementation of the simplicity constraint, which represents the difference between the full theory and its topological sector. The simplicity constraint is implemented as a suitable relation connecting the independent variables of the topological state-sum model. In the case of the spin-cube model, those variables are positive real numbers $L_\epsilon$ and integers $m_\Delta$, associated to edges and triangles in the triangulation, respectively. In the variables of the classical action (3.54) the simplicity constraint reads

$$B^{ab} = \varepsilon^{abcd} e_c \wedge e_d \,,$$

and is a classical equation of motion obtained by varying the action (3.54) with respect to $\phi_{ab}$. Passing to the variables $L_\epsilon, m_\Delta$ defined on the triangulation, the simplicity constraint takes the following form:

$$\gamma l_p^2 |m_\Delta| = A_H(L_{\epsilon_1}, L_{\epsilon_2}, L_{\epsilon_3}), \qquad \forall \Delta \in T(M), \tag{3.79}$$

where $\epsilon_1, \epsilon_2, \epsilon_3 \in \Delta$. Here $A_H(l)$ is the Heron formula for the area of the triangle $\Delta$ with the edge lengths $L_i$, given by

$$A_H(L_1, L_2, L_3) = \sqrt{s(s - L_1)(s - L_2)(s - L_3)}, \tag{3.80}$$

where $s \equiv (L_1 + L_2 + L_3)/2$ is the semiperimeter of the triangle, $\gamma$ is a dimensionless parameter determining the overall scale, while the unit of length is provided by the Planck length $l_p$. The simplicity constraint has a natural and obvious geometric interpretation — for every triangle $\Delta$ in $T$, the integer $m_\Delta$ determines the area of the triangle, with a universal proportionality constant $\gamma l_p^2$.

We should note that the factor $\gamma$ is a free parameter of the theory. It should not be confused with the Barbero-Immirzi parameter which is also commonly denoted by the same letter $\gamma$. The Barbero-Immirzi parameter can be introduced as a coupling constant if one adds the so-called Holst term [62] to the action (3.54). Since the Barbero-Immirzi parameter is not important for our purposes, we will not use it.

One can immediately see two basic properties of the simplicity constraint. First, it imposes triangle inequalities on all edge lengths $L_\epsilon$ in the triangulation. Namely, in triangulations with Lorentzian signature it is perfectly possible for the three edge lengths to violate triangle inequalities. This would give rise to an imaginary value of area (3.80), which corresponds to a triangle coplanar with a time axis in Minkowski spacetime. However, simplicity constraint (3.79) requires all triangles to be "space-like", i.e. to have real-valued (and positive) areas, which in turn imposes the triangle inequalities through (3.80). Second, the simplicity constraint transforms the Regge-like action from (3.50),

$$S_R(L, m) = \sum_{\Delta \in T} |m_\Delta| \delta_\Delta(L)$$

into the proper Regge action

$$S_R(L) = \frac{1}{\gamma l_p^2} \sum_{\Delta \in T} A_\Delta(L) \delta_\Delta(L) \,. \tag{3.81}$$

However, there is one big issue with the simplicity constraint (3.79). Namely, as it was first noted in [87], also see section 3.5.2, the system of equations (3.79) is not guaranteed to have any solutions. This can be seen as follows. If the total number of edges in $T$ is $E$, and the total number of triangles is $F$, then for every triangulation we have $F \geqslant E$, while the equality holds only for a single 4-simplex. This means that we have in total $F$ equations for $F$ integer variables and $E$ real variables. If we write the system generically as

$$|m_1| = A_1(L_1, \dots, L_E) \,,$$

$$\vdots$$

$$|m_E| = A_E(L_1, \dots, L_E) \,,$$
$$|m_{E+1}| = A_{E+1}(L_1, \dots, L_E) \,,$$

$$\vdots$$

$$|m_F| = A_F(L_1, \dots, L_E) \,,$$

where $A_1, \dots, A_F$ are suitable functions, we can in principle solve the first $E$ equations for $L_\epsilon$ as functions of $m_1, \dots, m_E$. Substituting those expressions into the remaining $F - E$ equations, the latter can be written in the form

$$|m_{E+1}| = f_1(m_1, \dots, m_E) \,,$$

$$\vdots \tag{3.82}$$

$$|m_F| = f_{F-E}(m_1, \dots, m_E) \,,$$

where $f_1, \ldots, f_{F-E}$ are implicitly defined functions, too complicated to be expressible in terms of elementary functions. Given that all $m_\Delta$ are integers, equations (3.82) are thus very complicated Diophantine equations, and they are not guaranteed to have any solutions at all. Therefore, short of providing the proof that the set of solutions for (3.82) is never empty, the simplicity constraint system (3.79) may be impossible to enforce as it stands.

In [87] an alternative strategy was proposed, also see section 3.5.3, and that is to implement the constraint "weakly", such that it holds only in the classical limit of the theory. In particular, it should be written as

$$|m_\Delta| = \left\lfloor \frac{1}{\gamma \ell_p^2} A_H(L_{\epsilon_1}, L_{\epsilon_2}, L_{\epsilon_3}) \right\rfloor, \qquad \forall \Delta \in T(M), \qquad (3.83)$$

where again $\epsilon_1, \epsilon_2, \epsilon_3 \in \Delta$. Here $\lfloor \ldots \rfloor$ represents the "floor" function, which returns the integer part of its argument. This definition of the simplicity constraint completely circumvents the issue of Diophantine equations. Moreover, as argued in [87], in the classical limit (defined as an asymptotic expansion when $m_\Delta, L_\epsilon \to \infty$) the difference between (3.79) and (3.83) is always a lower order correction, making the two systems of equations asymptotically equivalent.

In the formalism of the state-sum model, the implementation of the weak simplicity constraint (3.83) is given by the following choice of the edge and triangle amplitudes,

$$\mathcal{A}_\epsilon(L, m) = 1, \qquad \mathcal{A}_\Delta(L, m) = \chi \left( |m_\Delta| - \left\lfloor \frac{1}{\gamma \ell_p^2} A_H(L_{\epsilon \in \Delta}) \right\rfloor \right),$$

where $\chi(x)$ is an indicator function, having value 1 if $x = 0$ and is zero otherwise. Substituting this into the state sum

$$Z_T = \int_{\mathbb{R}_0^+} dL_1 \cdots \int_{\mathbb{R}_0^+} dL_E \sum_{m_1 \in \mathbb{Z}} \cdots \sum_{m_F \in \mathbb{Z}} \prod_{\epsilon \in T} \mathcal{A}_\epsilon(L, m) \prod_{\Delta \in T} \mathcal{A}_\Delta(L, m) \, e^{iS_R(L,m)},$$
$$(3.84)$$

one can immediately perform the summations over all $m_\Delta$, and the state sum reduces to the one defining the Regge quantum gravity model:

$$Z_T^{\text{weak}} \sim \int dL_1 \cdots dL_E \, e^{iS_R(L)}, \qquad (3.85)$$

where the domain of integration over edge lengths is a complicated subset of $(\mathbb{R}_0^+)^E$ due to the triangle inequalities imposed by the simplicity constraint. The action in the exponent is

$$S_R(L) \equiv \sum_{\Delta \in T} \left\lfloor \frac{1}{\gamma \ell_p^2} A_H(L_{\epsilon \in \Delta}) \right\rfloor \delta_\Delta(L), \qquad (3.86)$$

which is asymptotically equal to the proper Regge action (3.81) in the classical limit $L_\epsilon \to kL_\epsilon$, $k \to \infty$ (see Appendix C in [125] for proof). Note that the choice of the edge and triangle amplitudes defines the measure of the discretized path integral, and the product of these amplitudes over all edges and triangles can be identified with an appropriate Jacobian determinant.

Nevertheless, one could argue that imposing the simplicity constraint weakly might be unsatisfactory on various grounds, and it is legitimate to ask if the constraint can be imposed strongly [125]. Under the assumption that the Diophantine system discussed above has at least one solution for $L_\epsilon$, denoted $L_\epsilon = L_\epsilon(\bar{m})$, where the $\bar{m}$ denote the set of variables unconstrained by the equations, one can implement the strong simplicity constraint (3.79) with the following choice of edge and triangle amplitudes:

$$\mathcal{A}_\epsilon(L, m) = \delta\left(L_\epsilon - L_\epsilon(\bar{m})\right),$$
$$\mathcal{A}_\Delta(L, m) = \chi\left(|m_\Delta| - \frac{1}{\gamma l_p^2} A_H\left(L_{\epsilon \in \Delta}(\bar{m})\right)\right).$$

Substituting into (3.84), we can now integrate over all edge lengths, and evaluate all sums over $m$ except those over $\bar{m}$. The resulting state sum is

$$Z_T^{\text{strong}} \sim \sum_{\{\bar{m}\}} e^{iS_R(L(\bar{m}))}, \tag{3.87}$$

where the sum is taken over all independent sets of integers $\bar{m}$. The action in the exponent is

$$S_R(L(\bar{m})) \equiv \frac{1}{\gamma l_p^2} \sum_{\Delta \in T} A_H(L_{\epsilon \in \Delta}(\bar{m})) \delta_\Delta(L(\bar{m})),$$

and it is equal to the proper Regge action (3.81) evaluated on the simplicity constraint solution $L_\epsilon = L_\epsilon(\bar{m})$.

The implementation of the strong simplicity constraint rests upon the assumption that the complicated Diophantine equations (3.82) have a non-empty set of solutions, which is not obvious. It may thus come as a fresh surprise that a few classes of solutions can indeed be found. Moreover, the resulting spin-cube model can then be naturally related to a completely independent approach to quantum gravity, namely the Causal Dynamical Triangulations approach [3]. This is both completely unexpected and a very interesting result.

In the next section, we will perform an explicit construction of one class of exact solutions of the strongly imposed simplicity constraint, and after that we will discuss the relation between the spin-cube model and the CDT approach to quantum gravity.

### 3.5.5 *Solution of the simplicity constraint*

Let us turn to a constructive proof that the simplicity constraint (3.79) always has at least one solution. The proof will be done in three steps. First we will discuss the case of a single 4-simplex, then the case of two 4-simplices sharing a common tetrahedron, and finally the general case of an arbitrary triangulation.

A single 4-simplex has 10 edges and 10 triangles. In that case, the simplicity constraint (3.79) has the general form:

$$|m_1| = \frac{1}{\gamma l_p^2} A_1(L_1, \ldots, L_{10}),$$

$$\vdots \tag{3.88}$$

$$|m_{10}| = \frac{1}{\gamma l_p^2} A_{10}(L_1, \ldots, L_{10}).$$

This system of equations always has solutions. A simplest example is an equilateral 4-simplex, with areas and edges given as

$$m_1 = \cdots = m_{10} = k, \quad L_1 = \cdots = L_{10} = 2l_p \sqrt{\frac{\gamma |k|}{\sqrt{3}}},$$

for any $k \in \mathbb{Z}$. For more complicated choices of $m$'s, numerical analysis suggests that several solutions for $L$'s may exist, since there are 10 polynomial equations to be solved.

However, note that we require the edge lengths to be real-valued and positive, and moreover they must satisfy triangle inequalities. The choice of $m$'s might be such that this is impossible to satisfy, in which case the system does not have any solutions. Therefore, it is important to stress that we are not attempting to solve the system for ten $L$'s given any arbitrary $m$'s, but rather to solve the system for $L$'s and for $m$'s simultaneously, within their respective domains. The equilateral example proves that the set of solutions is non-empty, and numerical analysis (of Monte-Carlo type) shows that the set of solutions is actually quite rich (see Appendix D in [125] for details).

Next we move to a less trivial case of two 4-simplices sharing a common tetrahedron. The 4-dimensional figure is depicted on the following diagram:

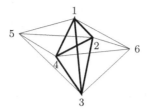

The first 4-simplex, $\sigma_1$, is determined by the vertices $(1, 2, 3, 4, 5)$, while the second, $\sigma_2$, is determined by the vertices $(1, 2, 3, 4, 6)$. They share the common tetrahedron $\tau$ determined by the vertices $(1, 2, 3, 4)$, depicted with thick edges. There are four triangles of $\tau$, shared by both 4-simplices, namely

$$(1, 2, 3), \quad (1, 2, 4), \quad (1, 3, 4), \quad (2, 3, 4).$$

In addition, $\sigma_1$ contains six more triangles

$$(5, 1, 2), \quad (5, 1, 3), \quad (5, 1, 4), \quad (5, 2, 3), \quad (5, 2, 4), \quad (5, 3, 4),$$

while $\sigma_2$ contains its own additional six triangles

$$(6, 1, 2), \quad (6, 1, 3), \quad (6, 1, 4), \quad (6, 2, 3), \quad (6, 2, 4), \quad (6, 3, 4).$$

In total, the figure has $E = 14$ edges and $F = 16$ triangles. For each triangle we write a simplicity constraint equation, giving rise to 16 equations:

$$|m_{123}| = \frac{1}{\gamma l_p^2} A_H(L_{12}, L_{13}, L_{23}),$$

$$\vdots \tag{3.89}$$

$$|m_{634}| = \frac{1}{\gamma l_p^2} A_H(L_{63}, L_{64}, L_{34}).$$

As discussed in the previous case, one can always solve the system of the first 10 equations, for the 4-simplex $\sigma_1$. Denote this solution as

$$m_{123} = \bar{m}_{123}, \qquad L_{12} = L_{12}(\bar{m}),$$

$$\vdots \qquad\qquad \vdots \tag{3.90}$$

$$m_{534} = \bar{m}_{534}, \qquad L_{54} = L_{54}(\bar{m}).$$

One could now arguably also solve four of the remaining six equations, for example for triangles $(6, 1, 2)$, $(6, 1, 3)$, $(6, 1, 4)$ and $(6, 2, 3)$, giving rise to additional four $\bar{m}$'s and four $L$'s. Substituting all previous results into the remaining two equations, one gets

$$|m_{624}| = f(\bar{m}, L(\bar{m})), \qquad |m_{634}| = g(\bar{m}, L(\bar{m})). \tag{3.91}$$

Here $f$ and $g$ are two functions implicitly defined by the substitution of 14 $\bar{m}$'s and 14 $L$'s into the remaining two equations. However, as discussed previously, there is *a priori* no guarantee that there will exist a choice for $(\bar{m}, L(\bar{m}))$ such that $m_{624}$ and $m_{634}$ are integers. But there is a beautiful geometrical argument which proves that this choice indeed exists. Namely, given the generic solution (3.90) for $\sigma_1$, choose the following for $\sigma_2$:

$$L_{61} = L_{51}(\bar{m}), \quad L_{62} = L_{52}(\bar{m}), \quad L_{63} = L_{53}(\bar{m}), \quad L_{64} = L_{54}(\bar{m}).$$
$$(3.92)$$

In other words, we choose the edges of $\sigma_2$ to be equal to corresponding edges of $\sigma_1$, making the two 4-simplices identical (up to reflection symmetry). Since the triangle areas of $\sigma_1$ are integers by construction, so will be the triangle areas of $\sigma_2$. In particular, we have

$$
\begin{aligned}
m_{612} &= \bar{m}_{512}, & m_{623} &= \bar{m}_{523}, \\
m_{613} &= \bar{m}_{513}, & m_{624} &= \bar{m}_{524}, \\
m_{614} &= \bar{m}_{514}, & m_{634} &= \bar{m}_{534}.
\end{aligned}
\qquad (3.93)
$$

In this way, all 16 simplicity constraint equations are simultaneously satisfied, giving us one explicit solution for the case of two connected 4-simplices. Numerical search for other solutions of the two equations (3.91) has proved fruitless, suggesting that this is the only possible generic solution. In special cases (like isosceles 4-simplices) there may be more solutions, and examples of these will be given in the next section.

Finally, having studied the special cases of one and two 4-simplices, we now turn to the general case of a triangulation containing arbitrary many simplices. The proof of the existence of a solution to the simplicity constraint (3.79) is simple, and is performed by induction over the number of 4-simplices. As a base case, note that a single simplex can be labeled with 10 $\bar{m}$'s and 10 $L$'s such that the simplicity constraint is satisfied, as already discussed above. To prove the inductive step, first assume that we have found a solution $(\bar{m}, L)$ that labels $N$ 4-simplices. Then, assuming that one of those 4-simplices features a free boundary tetrahedron, we proceed to attach the $(N + 1)$-th 4-simplex to that boundary tetrahedron. We then specify the labels $(m, L)$ of that additional 4-simplex to be equal to the corresponding labels of its neighbor, by the recipe discussed in the case of two 4-simplices above (namely (3.92) and (3.93)). The analysis of two 4-simplices above then guarantees that those labels also satisfy the additional six simplicity constraint equations, since the equations are the same for the two 4-simplices, and they are satisfied for one of them by induction hypothesis. Therefore, we have constructed an extended set of labels $(\bar{m}, L)$ which satisfies the simplicity constraint for $N + 1$ 4-simplices. This completes the proof of the induction step.

At this point a few comments are in order. First, numerical search for solutions distinct from (3.92), (3.93) has failed to find any, suggesting that this solution of the simplicity constraint equations is the only possible one, at least in the generic case. Second, as we can see from the construction, once we choose some particular labeling for the first 4-simplex, all other 4-simplices in the triangulation are labeled in the same way, i.e. the triangulation consists of 4-simplices which are all mutually identical. This has rather non-trivial physical consequences, because the state sum (3.87) reduces to the form

$$Z_T^{\text{strong}} = \sum_{\bar{m}_1 \in \mathbb{Z}} \cdots \sum_{\bar{m}_{10} \in \mathbb{Z}} \sum_\alpha e^{iS_R(L_\alpha(\bar{m}))}, \qquad (3.94)$$

where $\alpha$ counts the number of possible solutions for $L$'s, given a specific set of $\bar{m}$'s, in the system (3.88). From the state sum, one can see that there are only 10 degrees of freedom in the whole spacetime, specified by the identical labeling of all 4-simplices in $T$. This is unsatisfactory from the physical point of view, because we expect that the theory gives general relativity in the classical limit, and general relativity has two physical degrees of freedom (i.e. two graviton polarizations) per every point in space, which is certainly more than 10 in total. This problem is commonly resolved by passing to the formalism of "second quantization", i.e. by taking an additional sum over various possible triangulations:

$$Z^{\text{sq}} \equiv \sum_{T \in \mathcal{T}} Z_T^{\text{strong}} = \sum_{T \in \mathcal{T}} \sum_{\bar{m}_1 \in \mathbb{Z}} \cdots \sum_{\bar{m}_{10} \in \mathbb{Z}} \sum_\alpha e^{iS_R(L_\alpha(\bar{m}))}. \qquad (3.95)$$

The set $\mathcal{T}$ is some non-empty set of inequivalent triangulations $T(M_4)$, keeping the topology of the manifold $M_4$ the same. One can in principle attempt to discuss "all" possible triangulations, but it is often more useful to restrict to a certain class, according to various physical and mathematical criteria. Obviously, one of these criteria is the convergence of the state sum (3.95), assuming of course that (3.94) is finite to begin with.

We will not discuss the convergence issues or the choice of $\mathcal{T}$. We just note that since $\mathcal{T}$ can be chosen to be suitably large, it can provide an arbitrarily large number of degrees of freedom in the theory, through different configurations of the triangulation. It is however important to emphasize that this is a highly non-trivial step, and there are *a priori* no guarantees that summing over some set of triangulations will indeed introduce the needed degrees of freedom into the theory. It may still happen that the space of solutions is overconstrained.

As a final comment, note that the sums over $\bar{m}$'s and $\alpha$ are labeling all 4-simplices at the same time, independently of the choice of $T$. Under the implicit assumption of uniform convergence of the sum over triangulations, one is therefore allowed to switch the order of summations in (3.95) to obtain

$$Z^{\text{sq}} = \sum_{\bar{m}_1 \in \mathbb{Z}} \cdots \sum_{\bar{m}_{10} \in \mathbb{Z}} \sum_\alpha \left( \sum_{T \in \mathcal{T}} e^{iS_R(L_\alpha(\bar{m}))} \right) , \qquad (3.96)$$

where the term in parentheses becomes very similar to the state sum discussed in the models of CDT approach to quantum gravity. In section 6.2 we will analyze this relationship in more detail.

## 3.6   Regge path integrals

Let us now turn to the construction of a state-sum model for the Regge action (2.9). The corresponding Euclidean path integral can be written as

$$Z_E = \int_D \prod_{\epsilon=1}^{N_1} dL_\epsilon \, \mu(L) \, e^{-S_R(L)/l_P^2} , \qquad (3.97)$$

where $D$ is a subspace of $(\mathbb{R}_+)^{N_1}$ consistent with the triangular inequalities and $S_R(L)$ is the Regge action of length square dimension. One can introduce the path-integral measure $\mu(L)$ as

$$\mu(L) = \prod_{\epsilon=1}^{N_1} (L_\epsilon)^\alpha , \qquad (3.98)$$

where $\alpha$ is a constant, and usually $\alpha = 1$, see [59].

The immediate problem with the choices (3.97) and (3.98) is that the finiteness of $Z_E$ is not guaranteed because $S_R(L)$ is not bounded from below[4] and the measure (3.98) does not fall off sufficiently quickly for large $L_\epsilon$ and negative $\alpha$. A simple way to remedy this is to complexify the Euclidean path integral via

$$Z_{EC} = \int_D \prod_{\epsilon=1}^{N_1} dL_\epsilon \, \mu(L) \, e^{iS_R(L)/l_P^2} . \qquad (3.99)$$

---

[4]This is because the curvature scalar is not a bounded function of the metric, and this function can take values of any sign.

However, the problem with (3.99) is that it is not clear how to relate it to a path integral for the Minkowski signature metrics, i.e. how to do the Wick rotation.

In the case of the Minkowski signature PL metrics, the definition of the path integral requires some extra assumptions. Since the edge-lengths can be positive, zero or imaginary, one can restrict to only positive ones, i.e. the space-like edge lengths. However, for the purposes of defining an effective action, it is more natural and easier to use both space-like and time-like edge lengths, and this can be done in the following way. Let $M = \Sigma \times [0, n]$, $n \in \mathbb{N}$, be a spacetime manifold with two boundaries diffeomorphic to a 3-manifold $\Sigma$. We will use a time-ordered triangulation, which is also known as a causal triangulation [3]

$$T(M) = \bigcup_{k=0}^{n-1} \tilde{T}_k \left( \Sigma \times [k, k+1] \right) , \qquad (3.100)$$

where $\tilde{T}_k$ is a triangulation of a slab $\Sigma \times [k, k+1]$ such that

$$\partial \tilde{T}_k = T_k(\Sigma) \cup T_{k+1}(\Sigma)$$

and $T_k$ are triangulations of $\Sigma$. We then choose $v_\epsilon = L_\epsilon$ for $\epsilon \in T_k(\Sigma)$ and $v_\epsilon = iL_\epsilon$ for $\epsilon \in \tilde{T}_k \setminus (T_k \cup T_{k+1})$, see section 2.1.1 for the definition of $v_\epsilon$.

The corresponding path integral can be defined as

$$Z_R = \int_D \prod_{\epsilon=1}^{N_1} dL_\epsilon \, \mu(L) \, e^{i\tilde{S}_R(L)/l_P^2} , \qquad (3.101)$$

where $D$ is a region of $(\mathbb{R}_+)^{N_1}$ consistent with a PL Minkowski geometry[5] and $\tilde{S}_R$ is given by (2.26). The path-integral measure $\mu(L)$ must be chosen such that it makes $Z_R$ convergent. Furthermore, if we want a quantum effective action $\Gamma(L)$, which can be associated to (3.101), to become $\tilde{S}_R(L)$ in the classical limit ($L_\epsilon \gg l_P$), then the measure $\mu(L)$ has to obey

$$\ln \mu(\lambda L_1, \dots, \lambda L_N) \approx O(\lambda^a) , \qquad a \geqslant 2 , \qquad (3.102)$$

for $\lambda \to +\infty$, see [88, 99] and section 4.5.

A choice for $\mu(L)$ can be made such that it is consistent with (3.102) and it ensures the diffeomorphism invariance of the smooth-spacetime effective action [88]. It is given by

$$\mu(L) = \exp\left( -V_4(L)/L_0^4 \right) , \qquad (3.103)$$

where $V_4$ is the volume of $T(M)$ and $L_0$ is a new parameter in the theory. A set of possible values for $L_0$ can be fixed by requiring that the effective cosmological constant coincides with the observed value, see [99, 100] and section 5.1.

---

[5]The easiest way to determine $D$ is to embed $T(M)$ into a higher-dimensional Minkowski space.

### 3.6.1  *Finiteness of the Regge path integral*

One expects that the measure (3.103) guarantees the convergence of the path integral, because the function (3.103) vanishes exponentially fast in the regions where some or all $L_\epsilon \to \infty$. However, this expectation is not correct, because there are degenerate configurations such that $V_4(L) \to 0$ although some or all $L_\epsilon \to \infty$.

Let us denote the union of such regions as $D_0$ and let $D' = D \setminus D_0$, then

$$|Z| \leqslant \int_D \prod_\epsilon dL_\epsilon \, e^{-V_4(L)/L_0^4} \approx \int_{D_0} \prod_\epsilon dL_\epsilon + \int_{D'} \prod_\epsilon dL_\epsilon \, e^{-V_4(L)/L_0^4} \,.$$

$$(3.104)$$

The integral over $D_0$ is clearly divergent, so that we cannot prove the absolute convergence of $Z$. This does not mean that $Z$ is divergent, but it is an indication that we have to check the convergence of $Z$ in some other way.

By using the spherical coordinates in $\mathbb{R}_+^N$, the path integral $Z$ can be written as

$$Z = c_N \int_0^\infty L^{N-1} \, dL \int_{\Omega_N} d^{N-1}\theta \, J_N(\theta) \, e^{-L^4 v(\theta)/L_0^4 + iL^2 s(\theta)/l_P^2} \,, \qquad (3.105)$$

where

$$L^2 = \sum_{\epsilon=1}^N L_\epsilon^2 \,,$$

and

$$\theta = (\theta_1, \theta_2, \ldots, \theta_{N-1}) \,,$$

are the angles corresponding to the points of $\mathbb{R}_+^N$. The functions $v$ and $s$ are given by $V_4/L^4$ and $S_R/L^2$, while $c_N$ is a constant and $J_N(\theta)$ denotes the angular dependence of the Jacobian.

If we integrate first the $L$ variable, we obtain

$$Z = c_N \int_\Omega d^{N-1}\theta \, J_N(\theta) \int_0^\infty L^{N-1} \, e^{-v(\theta)L^4/L_0^4 + is(\theta)L^2/l_P^2} \, dL$$

$$= \int_\Omega d^{N-1}\theta \, J_N(\theta) \, F(v(\theta), s(\theta)) \,, \qquad (3.106)$$

where $\Omega$ is the angular region corresponding to the path-integral integration region $D \subset \mathbb{R}_+^N$. The function $F(v, s)$ has the following asymptotic properties. For $s \to 0$ we have

$$F(v, s) \approx c_N \int_0^\infty L^{N-1} \, e^{-vL^4/L_0^4} \, dL$$

$$= c_N v^{-N/4} \int_0^\infty \xi^{N-1} \, e^{-\xi^4/L_0^4} \, d\xi = c_N' \, v^{-N/4} \,. \qquad (3.107)$$

Similarly, for $v \to 0$ we obtain

$$F(v,s) \approx c_N \int_0^\infty L^{N-1} e^{-sL^2/l_P^2} \, dL = c_N'' \, s^{-N/2} . \tag{3.108}$$

Since $\Omega$ is a compact set, the angular integral (3.106) will diverge if $F$ is not bounded in $\Omega$. From (3.107) and (3.108) we can see that $F$ is not bounded when $v \approx 0$ and $s \approx 0$. Note that $v \approx 0$ corresponds to $V_4(L) \approx 0$, and $s \approx 0$ corresponds to $S_R(L) \approx 0$. Hence the integral $Z$ with the measure (3.103) is divergent.

Let us now consider the following class of PI measures

$$\mu_p(L) = \prod_\epsilon (l_0^2 + L_\epsilon^2)^{-p} \exp\left(-V_4(L)/L_0^4\right) , \tag{3.109}$$

such that $p > 1/2$ and $l_0 > 0$. We have a product of the edge-length measures which multiplies the 4-volume measure (3.103), and the new factor is essentially the same as the product of the edge-length measures (3.77). We than have

$$|Z| \leqslant \int_D \prod_\epsilon \frac{dL_\epsilon}{(l_0^2 + L_\epsilon^2)^p} \, e^{-V_4(L)/L_0^4} \approx \int_{D_0} \prod_\epsilon \frac{dL_\epsilon}{(l_0^2 + L_\epsilon^2)^p} + \int_{D'} d^N L \, \mu_p(L) .$$

The first integral is convergent for $p > 1/2$, while the second integral is also convergent, and this can be shown by using the $N$-dimensional spherical coordinates in $D'$. We can write

$$|Z| \leqslant I_2 = C + c' \int_{L'}^\infty L^{N-1} \, dL \int_{\Omega'} d^{N-1}\theta J_N(\theta) \, \frac{e^{-L^4 v(\theta)/L_0^4}}{L^{2Np}} ,$$

where $L' \gg l_0$. Consequently

$$I_2 \approx C + c' \int_{\Omega'} d^{N-1}\theta J_N(\theta) \, (v(\theta))^{N(p-1/2)/2} \int_{\xi'}^\infty \xi^{N-1} e^{-\xi^4/L_0^4} \, d\xi$$

$$= C + c'' \int_{\Omega'} d^{N-1}\theta J_N(\theta) \, (v(\theta))^{\frac{N}{2}(p-\frac{1}{2})} .$$

The angular integral in $I_2$ is finite, because it is an integral of a bounded function over a compact region. Hence $Z$ with the measure (3.109) is absolutely convergent. This measure also satisfies the criterion (3.102) for the construction of the effective action, so that it can be used to define a PL theory of quantum gravity.

One wonders whether the $p = 0$ measure could be still used, in spite of the fact that $Z$ is divergent in that case. We will see in the next section that the construction of the perturbative effective action does not depend on the finiteness of $Z$ and the relevant factor is a certain asymptotic behavior of the measure, which will be explained in the next section. This condition is satisfied for both the $p = 0$ and $p \neq 0$ measures. However, if one wants to construct a non-perturbative effective action, then the finiteness of $Z$ is important and one has to use a $p > 1/2$ measure.

## Chapter 4

# Effective actions for state-sum models

## 4.1 Effective action formalism

The effective action method originated in quantum field theory (QFT), and the idea was to find an effective field configuration which obeys field equations which can be considered as quantum corrected classical field equations, see [110]. These effective field configurations describe a QFT state evolution in the sense that those configurations correspond to the expectation value of the field operators in coherent states, see [35]. Since the corresponding field equations can be derived from an effective action, this gives the name for the corresponding method.

Note that the idea of effective trajectories and quantum corrected equations of motion (EoM) is also used in the De Broglie-Bohm formulation of quantum mechanics (QM), where the trajectories of particles are determined by the wavefunction which obeys the Schrödinger equation [25, 61]. One can show that the particle trajectories satisfy a quantum corrected Newton equation, where the quantum potential is determined by the wavefunction.

The effective action (EA) is also the generating function for the one-particle irreducible Feynman diagrams, see [110], and this property can be used to calculate the effective action perturbatively. However, one can easily derive the equation which determines the EA [101]. In the case of a scalar field in a flat Minkowski spacetime, with a classical action $S(\phi)$, see equation (4.7), the effective action is defined as

$$\Gamma(\phi) = W(J) - \int_{\mathbb{R}^4} d^4x J(x)\phi(x), \qquad (4.1)$$

where $J(x)$ is a solution of

$$\phi(x) = \frac{\delta W}{\delta J(x)}, \quad W(J) = -i\hbar \log Z(J),$$

and

$$Z(J) = \int \mathcal{D}\varphi \exp\left(\frac{i}{\hbar}\left(S(\varphi) + \int_{\mathbb{R}^4} d^4x J(x)\varphi(x)\right)\right).$$

The equation (4.1) implies

$$e^{\frac{i}{\hbar}\Gamma(\phi)} = \int \mathcal{D}\varphi \exp\left(\frac{i}{\hbar}S(\varphi) + \frac{i}{\hbar}\int_{\mathbb{R}^4} d^4x\, J(x)\left(\varphi(x) - \phi(x)\right)\right).$$

By changing the integration variable $\varphi(x)$ to $h(x) = \varphi(x) - \phi(x)$, we obtain

$$e^{\frac{i}{\hbar}\Gamma(\phi)} = \int \mathcal{D}h \exp\left(\frac{i}{\hbar}S(\phi + h) - \frac{i}{\hbar}\int_{\mathbb{R}^4} d^4x \frac{\delta\Gamma}{\delta\phi(x)} h(x)\right), \qquad (4.2)$$

which is the fundamental EA equation.

The equation (4.2) is an integro-differential equation, and by solving it, one could obtain an exact solution, which would give the non-perturbative effective action. However, the exact solutions of (4.2) are not known,[1] but the EA equation can be solved perturbatively by using

$$\Gamma = S + \hbar\Gamma_1 + \hbar^2\Gamma_2 + \cdots, \qquad (4.3)$$

see [69].

The equation (4.2) can also serve in the case of systems with finitely many degrees of freedom, i.e. the QM case. Since QM can be considered as a 1d QFT, we can write

$$e^{\frac{i}{\hbar}\Gamma(q)} = \int \mathcal{D}h \exp\left(\frac{i}{\hbar}S(q + h) - \frac{i}{\hbar}\int_a^b dt \frac{\delta\Gamma}{\delta q(t)} h(t)\right), \qquad (4.4)$$

where

$$S(q) = \int_a^b L(q, \dot{q})dt.$$

In the QM case, the EoM for the effective action defined by (4.4) have perturbative solutions which correspond to the expectation value of the Heisenberg operator $\hat{q}(t)$ in a coherent state $|(p_0, q_0)\rangle$, where $(p_0, q_0)$ are the initial values of the canonical momentum and the canonical coordinate, see [35].

In the case of the state-sum (SS) models, we can use the equation (4.4) if we consider an SS model as a dynamical system in a discrete set of time instants. Then the path integral in (4.4) becomes an ordinary integral, while the integrals over the time interval become sums. Equivalently, the space-time label $x$ becomes an element of a finite subset of $\mathbb{N}$, which corresponds

---

[1] An exception are free field theories, i.e. classical actions $S(\phi)$ which are quadratic in $\phi$. In those cases, the solution of (4.2) is $\Gamma(\phi) = S(\phi) + \text{const}$.

to the set of simplices of a triangulation and $q_k$ is a numerical value (edge length, triangle area, etc.) associated to a simplex $s_k$ of a triangulation. Hence

$$S(q) = f(q_1, q_2, \ldots, q_n) \, ,$$

so that

$$e^{\frac{i}{\hbar} \Gamma(q)} = \int_D \prod_k dh_k \, \exp\left( \frac{i}{\hbar} S(q+h) - \frac{i}{\hbar} \sum_k \frac{\partial \Gamma}{\partial q_k} h_k \right) , \qquad (4.5)$$

if $q$'s are continuous. The integration region $D \subseteq \mathbb{R}^n$ is determined by the condition $q + h \in Q$ for any $q \in Q$ where $Q \subseteq \mathbb{R}^n$ is the set of $q$-values. If $q$'s are discrete, then

$$e^{\frac{i}{\hbar} \Gamma(q)} = \sum_{h_1, \ldots, h_n} \exp\left( \frac{i}{\hbar} S(q+h) - \frac{i}{\hbar} \sum_k \frac{\partial \Gamma}{\partial q_k} h_k \right) , \qquad (4.6)$$

where $h \in D$.

Because the time variable disappears in the state-sum case, it is not clear what is the connection of a generic state-sum EA to the standard QM. However, the state-sum models discussed in this book can be thought of as discretizations of certain QFTs, i.e. as path integrals of discrete QFTs which are defined on a spacetime lattice or on a simplicial complex, which simplifies the task. However, the problem of time in QG makes this task more difficult, although there are some approaches in quantum cosmology, see section 5.2, which can circumvent this problem.

## 4.2   Wick rotation and QG

Because the EA equation has imaginary coefficients $i/\hbar$, a generic solution for $\Gamma$ will be complex. Hence we need some prescription in order to restrict a solution to real values. In the case of a QFT this is known as the Wick rotation.

We can see how this mechanism works in the case of a scalar field. Given the scalar field action

$$S = \int_{\mathbb{R}^4} d^3x \, dt \left[ \frac{1}{2} \left( \frac{\partial \phi}{\partial t} \right)^2 - \frac{1}{2} (\nabla \phi)^2 - U(\phi) \right] , \qquad (4.7)$$

then by allowing the time coordinate to take imaginary values, i.e. $t = -i\tau$, we obtain

$$S = i \int_{\mathbb{R}^4} d^3x \, d\tau \left[ \frac{1}{2} \left( \frac{\partial \phi}{\partial \tau} \right)^2 + \frac{1}{2} (\nabla \phi)^2 + U(\phi) \right] = iS_E \,.$$

$S_E$ is known as the Euclidean action, since it can be considered as the scalar field action on a 4-dimensional Euclidean space, i.e. $\mathbb{R}^4$ with the metric

$$ds^2 = dx^2 + dy^2 + dz^2 + d\tau^2 \,.$$

Consequently

$$e^{iS/\hbar} = e^{-S_E/\hbar} \,,$$

so that the EA equation for a scalar field on a Euclidean space becomes real, i.e.

$$e^{-\Gamma_E(\phi)/\hbar} = \int \mathcal{D}h \exp \left( -\frac{1}{\hbar} S_E(\phi + h) + \frac{1}{\hbar} \int_{\mathbb{R}^4} d^4x \, \frac{\delta \Gamma_E}{\delta \phi(x)} h(x) \right) \,. \quad (4.8)$$

Given an Euclidean solution $\Gamma_E(\phi)$, one obtains the Minkowski space solution by substituting the Euclidean metric in $\Gamma_E$ with the Minkowski metric, i.e.

$$\Gamma(\phi) = \Gamma_E(\phi) \Big|_{g_{\mu\nu} = \eta_{\mu\nu}} . \quad (4.9)$$

However, the Wick rotation cannot be applied to GR, because the Euclidean GR action is not positive definite, so that the path integral

$$Z(J) = \int \mathcal{D}g \, e^{-\left( S_E(g) + \int_M J(x)g(x) \, d^4x \right)/\hbar}$$

is not well-defined and consequently the corresponding EA equation will not be well-defined. Even if we ignore this problem, and solve the equation (4.8) perturbatively as

$$\Gamma_E(g) = S_E(g) + \hbar \Gamma_{1E}(g) + \hbar^2 \Gamma_{2E}(g) + \cdots \,,$$

the corrections $\Gamma_{kE}(g)$ will not be well-defined, because GR is perturbatively divergent and a non-renormalizable QFT.

One way to solve this problem is to use the following property of the effective action from QFT. Namely, at the one-loop level, i.e. when we use the approximation $\Gamma \approx S + \hbar \Gamma_1$, the Wick rotation (4.9) is equivalent to

$$\Gamma \to Re\,\Gamma + Im\,\Gamma \,, \quad (4.10)$$

since

$$\Gamma_1 = \frac{i}{2} \left[ \int_{\mathbb{R}^4} dx \int_{\mathbb{R}^4} dy \log \frac{\delta^2 S}{\delta \phi(x) \delta \phi(y)} \right]_{reg} \equiv \frac{i}{2} Tr \left( \log S''(\phi) \right) \,,$$

where $[\ldots]_{reg}$ stand for a regularized expression.

The definition (4.10) is suitable for quantum gravity, because it does not rely on the Euclidean QG. From the point of view of evaluating a path integral, it is better to use a Lorentzian path integral than the Euclidean one, because the integral of $e^{iS(g)/\hbar}$ has better convergence properties than the integral of $e^{-S(g)/\hbar}$. This can be seen by considering the following toy examples:

$$I_L = \int_0^\infty e^{iax^2}\, dx = e^{i\frac{\pi}{4}\,\mathrm{sgn}(a)}\frac{1}{2}\sqrt{\frac{\pi}{|a|}}\,,$$

and

$$I_E = \int_0^\infty e^{-ax^2}\, dx = \frac{1}{2}\sqrt{\frac{\pi}{a}}\,,$$

where $a \in \mathbb{R}$. Namely, $I_L$ is known as the Fresnel integral. It is a typical integral in the Lorentzian case, and it is convergent for any $a \neq 0$. On the other hand, $I_E$ is known as the Gaussian integral, and it is the Euclidean analog of $I_L$, but it is convergent only for $a > 0$.

## 4.3 EA for spin-foam models

As we explained in chapter 3, spin-foam models of quantum gravity in $D = 3$ and $D = 4$ spacetime dimensions can be described by a partition function of the form

$$Z(\Gamma) = \sum_{j,\iota}\prod_f W_2(j_f)\prod_l W_1(\iota_l)\prod_v W_0(j_{f(v)}, \iota_{l(v)})\,, \tag{4.11}$$

where $j = (j_1,\ldots,j_f,\ldots,j_F)$ label the faces of a 2-complex $\Gamma$ while $\iota = (\iota_1,\ldots,\iota_l,\ldots,\iota_L)$ label the edges of $\Gamma$. The $j$-labels correspond to a set of irreps of $Spin(D)$,[2] while $\iota$'s are the corresponding intertwiners. The two-complex $\Gamma$ is a subcomplex of the dual simplicial complex $T^*(M)$, where $T(M)$ is obtained by triangulating the spacetime manifold $M$. The weights $W_i$ are determined by the classical theory which we want to quantize, see chapter 3. Since $Z$ is just a complex number, it is not possible to extract the classical limit from it and one needs to analyze the boundary wavefunctions or to analyze the effective action.

A boundary wavefunction $\Psi_\Gamma(\hat\gamma)$ is associated to a boundary spin network $\hat\gamma = (\gamma, j_b, \iota_b)$, where $\gamma$ is the boundary one-complex of $\Gamma$ and $(j, \iota)$ are the corresponding labels. It corresponds to the situation when $M$ has a

---

[2]In the Euclidean gravity case, $Spin(3) = SU(2)$ and $Spin(4) = SU(2) \times SU(2)$, while in the Lorentzian case, $Spin(3) = SL(2, \mathbb{R})$ and $Spin(4) = SL(2, \mathbb{C})$.

boundary $\Sigma$, and $\Psi(\hat{\gamma})$ is constructed from (4.11) such that the summation is restricted to spin foams whose boundary spin network is $\hat{\gamma}$. Therefore

$$\Psi_\Gamma(\hat{\gamma}) = \sum_{j,\iota} \prod_f \tilde{W}_2(j_f) \prod_l \tilde{W}_1(\iota_l) \prod_v W_0(j_{f(v)}, \iota_{l(v)}), \qquad (4.12)$$

where the amplitudes $\tilde{W}$ are the same as the amplitudes $W$ for the faces and the links not belonging to the boundary, while for the boundary faces and links there is a choice which ensures good gluing properties, see [20].

Note that the construction (4.12) gives just one boundary state

$$|\Psi_\Gamma\rangle = \sum_{\hat{\gamma}} \Psi_\Gamma(\hat{\gamma})|\hat{\gamma}\rangle, \qquad (4.13)$$

for a given 2-complex $\Gamma$. However, we know from the canonical LQG that there are many more physical states. Especially important physical states are those which describe flat or constant curvature spatial manifolds. Therefore the definition (4.12) has to be changed such that the information about the background triads $E_0(x)$ is included, where $x$ is a spatial coordinate. In the case of Euclidean canonical LQG one can show that such a wavefunction has a form similar to (4.12), and the spin network $\hat{\gamma}$ is replaced by a spin network $\hat{\gamma}'$ where $\hat{\gamma}'$ is $\hat{\gamma}$ with the edge insertions $\mu_l(E_0(l))$, where $E_0(l) = \int_l E_0(x)dx$, see [85]. The insertion functions can be chosen freely, and an appropriate choice are the Gaussians centered around $E_0(l)$. In the case of a flat geometry, all $E_0(l)$ can be taken to be approximately the same, and the corresponding area of a triangle is proportional to $j_0$. This parameter will set the length scale, so that one introduces the insertions into the boundary spin network of (4.12) which will be functions of $j_0$.

Given such a $\Psi(\hat{\gamma}, j_0) \equiv \Psi_0(\hat{\gamma})$, there is the corresponding connection wavefunction $\Psi_0(A)$, which can be obtained by the loop transform. In order to see what is the SC limit which corresponds to this wavefunction, one uses

$$\Psi_0(A) = R(A)\, e^{i\mathcal{S}(A)/\hbar},$$

and has to consider the semiclassical expansion

$$\mathcal{S}(A) = \mathcal{S}_0(A) + \hbar\mathcal{S}_1(A) + O(\hbar^2).$$

If $\mathcal{S}_0(A)$ satisfies the Hamilton-Jacobi equation for canonical GR, then one can say that it corresponds to a physical wavefunction. It is obvious that this is an extremely difficult way to study the classical limit.

An easier approach would be to calculate the graviton correlation functions for the boundary state $\Psi_0(\hat{\gamma})$, so that

$$G_n(x_1, \ldots, x_n) = \sum_{\hat{\gamma}, \hat{\gamma}'} \Psi_0^*(\hat{\gamma}) \Psi_0(\hat{\gamma}') \langle \hat{\gamma} | \hat{h}(x_1) \cdots \hat{h}(x_n) | \hat{\gamma}' \rangle,$$

where $\hat{h}$ is the graviton operator. This was the approach started by Rovelli [114], and it can be shown that $G_2$ has the correct large-distance asymptotics if

$$\Psi_0(\hat{\gamma}) \approx N \exp \left( -\frac{1}{j_0} \sum_{l,l' \subset \gamma} \alpha_{ll'}(j_l - j_0)(j_{l'} - j_0) \right), \qquad (4.14)$$

for large spins, where $\alpha$ is a constant matrix [86].

However, the correct asymptotics of $G_2$ does not guarantee that the classical limit of a spin-foam model is general relativity. Namely, one has to show that all correlation functions $G_n$ correspond to the ones for the EH action in the classical limit. This is equivalent to demonstrating that the effective action

$$\Gamma[h] = \sum_{n \geqslant 2} c_n \int dx_1 dt_1 \cdots \int dx_n dt_n$$
$$\tilde{G}_n(x_1, t_1, \ldots, x_n, t_n) \, h(x_1, t_1) \cdots h(x_n, t_n), \qquad (4.15)$$

has the classical limit which is given by the Einstein-Hilbert action, where $(x, t)$ is a spacetime point coordinate and $\tilde{G}_n$ is the extension of $G_n$ when $t_k \neq t_l$. The correlation function approach is less difficult then the wave-function approach, but it still requires a lot of work.

A simpler method to compute the effective action would be a method where $\Gamma$ is given as a functional of the spacetime metric $g$ rather then the functional of $h = g - \eta$, where $\eta$ is the flat metric. The background field method (BFM) of computing the effective action in QFT [1], is a method convenient for this purpose. In the case of quantum gravity, the BFM approach suggests the following relation

$$e^{i\Gamma(g)/\hbar} \approx \int \mathcal{D}h \, e^{iS(g+h)/\hbar}, \qquad (4.16)$$

where $S(g)$ is the Einstein-Hilbert action and $S'(g) = 0$. From the exact formula for the effective action (4.2), we can see that (4.16) gives the one-loop approximation

$$\Gamma(g) \approx S(g) + \hbar \Gamma_1(g).$$

The expression (4.16) is formal and has to be defined, and in the perturbative QFT approach it amounts to fixing a gauge for the diffeomorphism gauge symmetry and implementing a regularization/renormalization procedure. This has to be done because the theory allows arbitrary short distances and hence the infinities appear. Since GR is non-renormalizable, the corresponding effective action cannot be determined uniquely.

In the case of spin-foam models, the problems with QFT infinities are avoided because the theory has a natural short distance cut-off. Namely, the basic degrees of freedom are the $SU(2)$ spins $j_f \in \mathbb{N}/2$, and these are essentially the areas of the corresponding triangles, since $A_f \propto l_P^2 \sqrt{j_f(j_f + 1)}$. Hence there is a short-distance cut-off of order of the Planck length $l_P$. There is no need for a gauge-fixing procedure, since $j_f$ are triangle areas, and these are diffeomorphism invariant. The only infinities which can appear are in the large $j_f$ region, which correspond to large-distance infinities (also known as the infra-red infinities in QFT), but these can be easily dealt with by introducing the appropriate negative powers of $j_f$ in the spin foam amplitude, see [97].

The formula (4.16) takes the following form in the case of spin-foam models

$$e^{i\Gamma(j,\iota)} \approx \sum_{j',\iota'} \prod_f W_2(j_f + j'_f) \prod_l W_1(\iota_l + \iota'_l) \prod_v W_0(j_{f(v)} + j'_{f(v)}, \iota_{l(v)} + \iota'_{l(v)}),$$

(4.17)

where $(j, \iota)$ is a configuration representing the background or the classical values of the spin foam labels, while the summation is over the fluctuations $(j', \iota')$ around the classical background $(j, \iota)$. The approximation (4.17) is valid if we take that the background spins are large and that $(j, \iota)$ is a stationary point of $Re\, S(j, \iota)$, where

$$e^{iS(j,\iota)} = \prod_f W_2(j_f) \prod_l W_1(\iota_l) \prod_v W_0(j_{f(v)}, \iota_{l(v)}) \qquad (4.18)$$

is the spin foam amplitude for $M = \Sigma \times I$ with two boundaries. This is the spin foam version of the BFM approximation for GR given by (4.16).

As discussed in the introduction, the vertex amplitude $W_0$ should have the asymptotics

$$W_0(j, \iota) \approx \frac{e^{i\alpha S_{vR}(j)}}{V_p(j)} \qquad (4.19)$$

for $j \to \infty$, where

$$S_{vR} = \sum_{f \supset v} j_f \theta_f \qquad (4.20)$$

is the vertex Regge action and $\theta_f$ are the dihedral angles, while $V_p(j)$ should be a homogeneous function of order $p > 0$. The role of the function $V_p(j)$ is to ensure that the state sum (4.17) is convergent, see [97].

The requirement (4.19) is essential, since only in this case, one can obtain the Regge action

$$S_R = l_P^2 \sum_f j_f \delta_f \tag{4.21}$$

as the classical limit of the effective action. Namely, if we take the background spins to be large, then we can use the asymptotic formula (4.19) in the state sum (4.17), which then produces the Regge action in the exponent due to the identity

$$S_R/l_P^2 = \sum_v S_{vR} + 2\pi \sum_f k_f j_f \,, \tag{4.22}$$

where $k_f$ are integers. It is important to notice that the deficit angle $\delta_f$ is given by

$$\delta_f = 2\pi - \sum_{v \subset f} \theta'_{fv} \,. \tag{4.23}$$

For a space-like face $f$ $\theta'_{fv} = \pi - \theta_{fv}$ is the interior dihedral angle for the 4-simplex $\sigma_v$ dual to $v$. A space-like $f$ means that the dual triangle $\Delta_f$ is time-like and belongs to $\sigma_v$. When $f$ is time-like, which means that $\Delta_f$ is space-like, then

$$\delta_f = \sum_{v \subset f} \Theta_{fv} \,, \tag{4.24}$$

where $\theta_{fv} = \Theta_{fv}$ and $\Theta_{fv}$ is the boost parameter between the normal vectors of the two tetrahedrons of $\sigma_v$ which share the triangle $\Delta_f$, see [14]. This definition is equivalent to the one given in section 2.1.

### 4.3.1 $D = 3$ case

Let us explore first the effective action given by (4.17) in the simpler case of three-dimensional (3d) spin-foam models. Consider the Ponzano-Regge model partition function for the Euclidean 3d gravity

$$Z_{PR} = \sum_j \prod_f (-1)^{2j_f} \dim j_f \prod_v W(j_{f(v)}) (-1)^{j_1(v)+\cdots+j_6(v)} \,, \tag{4.25}$$

where $W$ is the $6j$-symbol, see [15]. The immediate problem with $Z_{PR}$ is that it is divergent, so that it has to be regularized. This can be done by introducing a maximum spin, or by dividing $W$ by an appropriate power of the product of the dimensions of the vertex spins. The later regularization will be more convenient for our purposes. Either way one loses the triangulation independence, but we will see that this is not going to be a problem for our purposes.

The next problem is that

$$W(j) \approx \frac{\cos [S_{vR}(j)]}{\sqrt{V(j)}},$$

for large spins, where $V(j)$ is the volume of the vertex tetrahedron. According to our approach we then need to change the vertex amplitude such that the new asymptotics is given by (4.19). In order to achieve this, consider

$$\tilde{W} = \sqrt{V}W + \sqrt{VW^2 - 1}. \tag{4.26}$$

It is easy to show that $\tilde{W} \approx e^{iS_{vR}}$ for large spins since (4.26) implies

$$W = \frac{\tilde{W} + (\tilde{W})^{-1}}{2\sqrt{V}}.$$

Let us introduce a modified vertex amplitude

$$\mathcal{A}(j) = \frac{\tilde{W}(j)}{\sqrt{V} \prod_{k=1}^{6}(\dim j_k)^{p'}}. \tag{4.27}$$

Then $\mathcal{A}(j)$ will have the asymptotic form (4.19) with $p = 6p' + \frac{3}{2}$, when all of the six spins $j$ are large. The parameter $p'$ has to be chosen such that the state sum (4.17) is finite. This can be done because $\tilde{W}/\sqrt{V}$ is a bounded function. There exists $M > 0$ such that

$$|\mathcal{A}(j)| < \frac{M}{\prod_k(\dim j_k)^{p'}}. \tag{4.28}$$

Consequently

$$|Z_p| < N \sum_j \prod_f (\dim j_f)^{1-p'n_f} \leqslant N \sum_j \prod_f (\dim j_f)^{1-2p'}. \tag{4.29}$$

where $Z_p$ is the partition function associated to the state sum (4.17). The $n_f$ denotes the number of vertices of a face $f$ and since $n_f \geqslant 2$, we obtain the last inequality. The last sum in (4.29) will be convergent for $p' > 1$. Therefore $Z_p$ will be convergent for $p' > 1$. One can find a better estimate for the lower bound for $p'$ by using a better estimate for $\tilde{W}/\sqrt{V}$, but for us it is important that such $p'$ exist and that their values are triangulation-independent.

We are interested in calculating the effective action when all the background spins in the state sum (4.17) become large. Then one can approximate each vertex amplitude in (4.17) by using (4.19), since all the spin labels $j + j'$ are large. Consequently

$$e^{i\Gamma(j)/l_P} \approx N' \sum_{j'} \prod_f (j_f + j'_f)^{1-p_f n_f} e^{iS_R(j+j')/l_P}, \qquad (4.30)$$

where $p_f = p' + 3/2$,

$$S_R(j) = l_P \sum_{\epsilon \in T(M)} j_\epsilon \delta_\epsilon(j) = \sum_{\epsilon \in T(M)} l_\epsilon \delta_\epsilon(l), \qquad (4.31)$$

is the three-dimensional Regge action and $l_P = G_N \hbar$ is the three-dimensional Planck length. Note that the sign factors in the face and the vertex amplitudes, see (4.25), combine with the sum of the vertex Regge actions such that the Regge action is obtained in the exponent of (4.30).

The main contribution in the state sum (4.30) comes from $j'_f \ll j_f$, since the weights $(j_f + j'_f)^{1-p_f n_f}$ are maximal for $j'_f = 0$ and drop-off as negative powers of $j'_f$. We can then use

$$(j + j')^{-m} = j^{-m}\left(1 - \frac{j'}{j}\right)^{-m} = j^{-m}\left[1 - m\frac{j'}{j} + m(m+1)\frac{j'^2}{2j^2} + \cdots\right],$$

which is valid for $j' < j$. Consequently

$$e^{i\Gamma(j)/l_P} \approx N \sum_{j'} e^{iS_R(j+j')/l_P} \prod_f j_f^{-m_f}\left[1 - m_f\frac{j'_f}{j_f} + m_f(m_f+1)\frac{j'^2_f}{j^2_f}\right], \qquad (4.32)$$

where $m_f = n_f p_f - 1$ is a positive number.

Let us choose the background spins $j_f$ such that they correspond to a stationary point of the Regge action $S_R(j)$. This is a standard procedure in the case of QFTs, and the idea is to simplify the calculation, since the stationary point restriction of the background spins does not affect the important features of the effective action. Therefore we can use the approximation

$$S_R(j + j') \approx S_R(j) + \frac{l_P}{2}\sum_{f,f'} \tilde{S}''_{R\,ff'}(j)j'_f j'_{f'},$$

where $\tilde{S}''_{R\,ff'}(j)$ is the Hessian matrix for $S_R(j)/l_P$ and $S'_R(j) = 0$. Consequently

$$e^{i\Gamma(j)/l_P} \approx N\, e^{iS_R(j)/l_P - \sum_f m_f \ln j_f} \sum_{j'} e^{i\langle \tilde{S}''_R(j)j'j'\rangle/2} \prod_f \left(1 - m_f\frac{j'_f}{j_f} + \cdots\right), \qquad (4.33)$$

where $\langle \tilde{S}''_R(j)j'j' \rangle = \sum_{f,f'} \tilde{S}''_{Rff'}(j)j'_f j'_{f'}$. The sum in (4.33) can be approximated by an integral over $x_f = j'_f/j_f$ variables, and this integral will be given as a sum of the integrals of the following type

$$\int d^F x \, x_1^{n_1} \cdots x_F^{n_F} \exp\left(\frac{i}{2}\sum_{m,n} \tilde{S}''_{Rmn}x_m x_n\right).$$

These integrals can be calculated by taking the derivatives of the generating function

$$I(\mu) = \int d^F x \, \exp\left(\frac{i}{2}\sum_{m,n} \tilde{S}''_{Rmn}x_m x_n + \sum_m \mu_m x_m\right),$$

at $\mu = 0$, where $I(\mu)$ is given by

$$I(\mu) = (2\pi i)^{F/2} \frac{\exp\left(i\mu^T(\tilde{S}''_R)^{-1}\mu/2\right)}{\sqrt{\det(\tilde{S}''_R)}}, \tag{4.34}$$

and $\mu^T = (\mu_1, \ldots, \mu_F)$. This calculation can be simplified by using

$$(1+x)^{-m} = e^{-m\log(1+x)} = e^{-mx+mx^2/2+\cdots}$$

so that

$$e^{i\Gamma(j,\iota)/l_P} \approx N\, e^{-\sum_f m_f \log j_f + iS_R(j)/l_P} \sum_{j'} e^{-\sum_f m_f j'_f/j_f + \frac{i}{2}\sum_{f,f'} \tilde{S}''_{Rff'}j'_f j'_{f'}}, \tag{4.35}$$

where

$$\hat{S}''_{Rff'} = \tilde{S}''_{Rff'} - i\delta_{f,f'}\frac{m_f}{j_f^2} \approx \tilde{S}''_{Rff'},$$

for large $j_f$. Then by using (4.34)

$$e^{i\Gamma/l_P} \approx N' e^{\left(-\sum_f m_f \log j_f + iS_R(j)/l_P\right)} \frac{\exp\left(i\sum_{f,f'} m_f m_{f'}\frac{\tilde{G}_{ff'}(j)}{2j_f j_{f'}}\right)}{(\det \tilde{S}''_R(j))^{1/2}} \tag{4.36}$$

where $\tilde{G}$ is the inverse matrix of $\tilde{S}''$. Consequently

$$e^{i\Gamma/l_P} \approx N' \exp\left(-\sum_f m_f \log j_f + iS_R(j)/l_P \right.$$

$$\left. -\frac{1}{2}\mathrm{Tr}\log \tilde{S}''_R(j) + i\sum_{f,f'} m_f m_{f'}\frac{\tilde{G}_{ff'}(j)}{2j_f j_{f'}}\right). \tag{4.37}$$

The equation (4.37) implies that

$$\Gamma(j)/l_P \approx S_R(j)/l_P + i\sum_f m_f \log j_f + \frac{i}{2} Tr \log \tilde{S}''_R(j)$$

$$- i \log N' + \sum_{f,f'} m_f m_{f'} \frac{\tilde{G}_{ff'}(j)}{2 j_f j_{f'}}$$

$$\approx S_R(j)/l_P + i\sum_f m_f \log j_f + \frac{i}{2} Tr \log \tilde{S}''_R(j) - i \log N' + O(1/j),$$

$$(4.38)$$

where $O(1/j)$ denotes the terms which scale as $1/j$ when $j_f \to \infty$. When necessary, it will be understood that notation $O(j^m)$ also includes the subleading terms.

Note that $S_R = O(j)$, while $\log j_f$ and $Tr \log S''_R$ are of $O(\log j)$. Since an edge length $l_\epsilon$ is given by $l_P j_\epsilon$, the formula (4.38) implies

$$\Gamma(l) \approx S_R(l) + l_P \Gamma_1(l),$$

for $l \gg l_P$. The dominant term in the large-spin/large-length limit will the be Regge action (4.31). Hence we can say that the classical limit of the effective action is the Regge action, which is a discretization of the Einstein-Hilbert action. This means that if we start refining the spacetime simplicial complex, $S_R$ will become the EH action.

Note that the quantum correction terms in (4.38) are imaginary numbers, while the effective action should be a real function. The same problem appears in QFT, see section 4.2, since in that case, the one-loop approximation is given by

$$e^{i\Gamma(\phi)/\hbar} \approx \int \mathcal{D}\varphi \, e^{iS(\phi+\varphi)/\hbar}. \qquad (4.39)$$

The stationary phase approximation then implies

$$\Gamma(\phi) \approx S(\phi) + i\frac{\hbar}{2} Tr \log S''(\phi),$$

so that $\Gamma(\phi)$ is not a real action. This problem is resolved by resorting to the Wick rotation. Namely, by performing the Wick rotation $t \to it$, where $t$ is the time coordinate, one passes to the theory in the Euclidean metric and defines the effective action which is real, see section 4.2, while in the context of the one-loop approximation the relation (4.39) becomes

$$e^{-\Gamma(\phi)/\hbar} \approx \int \mathcal{D}\varphi \, e^{-S(\phi+\varphi)/\hbar}.$$

Consequently

$$\Gamma(\phi) \approx S(\phi) + \frac{\hbar}{2} Tr \log S''(\phi) \tag{4.40}$$

is real, and by replacing the Euclidean metric with the Minkowski metric in (4.40), one obtains a real effective action. In the case of spin-foam models, there are no spacetime coordinates and there is no background metric, so that one cannot perform the Wick rotation. Nevertheless, since the Wick rotation in QFT essentially amounts to a redefinition of a complex function $\Gamma = \Gamma_0 + i\hbar\Gamma_1$ into a real function $\Gamma_0 + \hbar\Gamma_1$, in quantum gravity we can define a real effective action by using the same prescription

$$\Gamma \to Re\,\Gamma + Im\,\Gamma. \tag{4.41}$$

The definition (4.41) is not unique, since one can also use $Re\,\Gamma - Im\,\Gamma$. This ambiguity can be only resolved by an experiment, but we will use (4.41), because it agrees with the QFT theory sign. Therefore for large spins we obtain

$$\Gamma(j) \approx S_R(j) + l_P \left( \sum_f m_f \log j_f + \frac{1}{2} Tr \log \tilde{S}''_R(j) + O(1/j) \right). \tag{4.42}$$

We would like to make the following remarks. The partition function $Z_p$, which corresponds to the modified vertex amplitude (4.27) is finite, but it is not a topological invariant. This is not a problem, since our goal is not constructing manifold invariants, but obtaining a quantum theory of gravity whose classical limit is general relativity. This is achieved by requiring that the state sum (4.25) is finite and that the classical limit of the effective action is the Regge action (4.31). Although our construction is triangulation dependent, it still leads to a topological theory in the continuum limit. Namely, if we start refining the triangulation, the Regge action will become the EH action, which defines a topological theory in three spacetime dimensions.

The trace-log term in (4.42) is a discretization of the usual trace-log term from QFT. Since the QFT trace-log term is divergent, the spin foam version can be considered as a regularization of the QFT counterpart. In contrast, the $m \log j$ terms in (4.42) do not have an analog in the QFT case, and their presence is a feature of the model. It is not clear what is the smooth limit of the $m \log j$ terms and whether their presence is a good or a bad feature of the model, since we do not know experimentally what are the quantum gravity corrections.

Note that one can define a spin-foam model with a simpler vertex amplitude than (4.27)

$$\tilde{A}(j) = \frac{e^{iS_{vR}(j)}}{\prod_k (\dim j_k)^p}.$$ (4.43)

The corresponding state sum would look like the path integral for a Regge gravity model where the lengths of the edges of a triangulation are positive half-integers. One can enforce the triangle inequalities by inserting the dual-edge (triangle) amplitudes proportional to the theta spin network evaluation, so that

$$\tilde{Z} = \sum_j \prod_f (-1)^{2j_f} \dim j_f \prod_l \theta(j_{f(l)}) \prod_v \tilde{A}(j_{f(v)})(-1)^{j_1(v)+\cdots+j_6(v)}.$$ (4.44)

The corresponding effective action will be also given by the expression (4.42).

The spin-foam model defined by the $\tilde{A}$ amplitude can be easily extended to the Lorentzian case, simply by replacing the vertex Regge action $S_{vR}$ in (4.43) by its Lorentzian analog. We will label the edges of the triangulation with the unitary irreps $j_f$ from the discrete series of representations of $Spin(1,2) = SL(2,R)$, which makes sense if the edges are space-like. Then the deficit angle is given by the 3d analog of (4.24), so that the sum of the vertex Regge actions will be equal to the Regge action. Therefore we will not need the sign factors in the face and vertex amplitudes, which were necessary in the Euclidean case, in order to obtain the Regge action in the classical limit of the effective action. Hence one can define a finite Lorentzian 3d quantum gravity spin-foam model whose partition function is given by

$$\tilde{Z}_L = \sum_j \prod_f (2j_f + 1) \prod_l \theta(j_{f(l)}) \prod_v \tilde{A}(j_{f(v)}).$$ (4.45)

The corresponding effective action is given by (4.42) in the large-spin limit so that in the smooth spacetime limit, one will obtain the EH action.

### 4.3.2  $D = 4$ case

We will now study the effective action for the EPRL/FK model. The exact QFT formula (4.2) can be applied to the EPRL/FK model as

$$e^{i\Gamma(j,\vec{n})/l_P^2} \approx \sum_{j'} \int d\vec{n}' \prod_f W_f(j+j') \prod_v W_v(j+j', \vec{n}+\vec{n}') e^{-\frac{i}{l_P^2}\left\langle \frac{\partial \Gamma}{\partial j} j' + \frac{\partial \Gamma}{\partial n} \vec{n}' \right\rangle},$$ (4.46)

where

$$\left\langle \frac{\partial \Gamma}{\partial j} j' + \frac{\partial \Gamma}{\partial \vec{n}} \vec{n}' \right\rangle = \sum_f \frac{\partial \Gamma}{\partial j_f} j'_f + \sum_{fl} \frac{\partial \Gamma}{\partial \vec{n}_{fl}} \vec{n}'_{fl},$$

is the spin foam analog of the higher-loop correction term from (4.2).

The background $(j, \vec{n})$ can be chosen to be a Regge background by choosing $(j(L), \vec{n}(L))$, where the lengths $L$ are off-shell, i.e. do not have to satisfy the Regge equations of motion. However, the EA method allows an arbitrary background, so we will not restrict our analysis to a Regge background.

The analog of the $\hbar$-expansion (4.3) can be written as

$$\Gamma(j, \vec{n}) = \Gamma_0(j, \vec{n}) + l_P^2 \Gamma_1(j, \vec{n}) + l_P^4 \Gamma_2(j, \vec{n}) + \ldots,$$

and on the basis of dimensional analysis, one must have

$$\Gamma_0 = O(j)l_0^2, \quad \Gamma_1 = O(\log j), \quad \Gamma_2 = O(j^{-2})l_0^{-2}, \ldots,$$

where $l_0$ is some unit of length. A spin-foam model will have the correct classical limit, if one can show that

$$\Gamma_0(j, \vec{n}) \approx S_R(L)$$

for $j_f \to \infty$, $f = 1, 2, \ldots, F$, and the $j$'s are such that the triangle areas $A_\Delta \approx j_f l_P^2$, where the triangle $\Delta$ is dual to a face $f$, can be expressed via the Heron formulas for the edge lengths $L_\epsilon$, $\epsilon = 1, 2, \ldots, E$. In the case when $j_f$ do not correspond to an assignment of the edge lengths, $\Gamma_0$ should be some generalization of the Regge action.

The EPRL/FK spin-foam model [44, 49] is a good candidate for a QG theory because the simplicity constraints are implemented, see section 3.2.1, and it has the correct spin-network labels on a three-dimensional boundary.

The partition function (4.11) is given by

$$W_2(j) = \dim j, \quad W_1(j, \iota) = 1$$

and

$$W_0(j, \iota) = \sum_{n_1 \geqslant 0, \cdots, n_5 \geqslant 0} \prod_{a=1}^{5} \int_0^{+\infty} d\rho_a (n_a^2 + \rho_a^2) \hat{f}_{n_a \rho_a}^{\iota_a}(j) \, W_{15}(2j_{bc}, 2\gamma j_{bc}; n_b, \rho_b),$$

where $\gamma$ is the Barbero-Immirzi parameter, $W_{15}$ is the $15j$-symbol for the unitary representations $(n, \rho)$ of the Lorentz group and $\hat{f}$ are the fusion coefficients, see [44] for the details. $W_2(j)$ was originally chosen to be a quadratic function [44], but it has been recently argued in [20] that the linear weight is more appropriate. In any case, the essential features of the effective action will be the same.

A more convenient form of $Z$ is

$$Z = \sum_{j,\vec{n}} \prod_f \dim j_f \prod_v W(j_{f(v)}, \vec{n}_{lf(v)}),$$

where each $\iota_l$ in a spin foam from the sum (4.11) is replaced by four unit 3-vectors $\vec{n}_{lf}$. An $\vec{n}_{lf}$ is a 3-vector orthogonal to the triangle dual to face $f$, such that this triangle belongs to the tetrahedron dual to a link $l$.

For the purposes of calculating the effective action we only need to now the asymptotics of $W(j, \vec{n})$ when all the spins $j$ are large, also see section 3.2.1. This asymptotics is given by

$$W(j, \vec{n}) \approx \frac{N_+(\alpha)e^{i\alpha S_{vR}(j,\vec{n})} + N_-(\alpha)e^{-i\alpha S_{vR}(j,\vec{n})}}{V(j)} \tag{4.47}$$

where $\alpha = \gamma$ when the 4-simplex boundary geometry is Lorentzian, while $\alpha = 1$ when the 4-simplex boundary geometry is Euclidean, see [13]. The real numbers $N_\pm(\alpha)$ are $O(1)$ functions of the spins, while $V = O(j^{12})$. There are also degenerate configurations of spins for which $W(j, \vec{n}) \approx N(j)/V(j)$ where $N(j) = O(1)$ and $V(j) = O(j^{12})$, but their contribution to the state sum is negligible. Otherwise, $W$ drops off faster than any power of $1/j$.

In order to obtain a correct $\Gamma_0$ we need to redefine the vertex amplitude. Let us introduce a new vertex amplitude $\tilde{W}(j, \vec{n})$ such that

$$\tilde{W} = \frac{VW + \sqrt{(VW)^2 - 4N_+N_-}}{2N_+}. \tag{4.48}$$

This formula follows from the following relation between $W$ and $\tilde{W}$

$$W = \frac{N_+\tilde{W} + N_-(\tilde{W})^{-1}}{V(j)},$$

which ensures that

$$\tilde{W}(j, \vec{n}) \approx e^{i\gamma S_{vR}(j,\vec{n})},$$

for large spins. One can then define

$$\mathcal{A}(j, \vec{n}) = \frac{\tilde{W}(j, \vec{n})}{V(j)\prod_f(\dim j_f)^p}, \tag{4.49}$$

where $p$ is sufficiently large such that $Z_p$ is convergent. Such a $p$, which does not depend on the triangulation, can be always arranged, as shown in [93].

Note that there is another way to modify the EPRL/FK vertex amplitude in order to obtain the desired single-exponent asymptotics, see [45]. In this approach one modifies the integral representation of the EPRL/FK vertex by introducing an extra factor in the integrand, the so-called "projector term".

The amplitude (4.49) has the desired asymptotics and it gives a finite partition function. We can now use the formula (4.17) for the effective action, which takes the form

$$e^{i\Gamma(j,\vec{n})/l_P^2} \approx \sum_{j',\vec{n}'} \prod_f [2(j_f + j'_f) + 1] \prod_v \mathcal{A}(j_{f(v)} + j'_{f(v)}, \vec{n}_{fl(v)} + \vec{n}'_{fl(v)}) \, .$$

$$(4.50)$$

The labeling of the EPRL/FK model is consistent with a space-like triangulation. In that case $k_f = 0$ in the formula (4.22), so that when $(j_1, \ldots, j_F) \to (\infty, \ldots, \infty)$ we obtain

$$e^{i\Gamma(j,\vec{n})/l_P^2} \approx N \sum_{j',\vec{n}'} \prod_f (j_f + j'_f) \prod_v \frac{e^{-i\gamma S_{vR}(j_{f(v)} + j'_{f(v)}, \vec{n}_{fl(v)} + \vec{n}'_{fl(v)})}}{V(j) \prod_f (j_f + j'_f)^p}$$

$$\approx N \sum_{j',\vec{n}'} \prod_f (j_f + j'_f)^{1-p_f m_f} e^{i\gamma S_R(j+j', \vec{n}+\vec{n}')/l_P^2} \, , \qquad (4.51)$$

where $p_f \geqslant p$ and $m_f$ is the number of vertices belonging to a face $f$. Note that the contribution to $e^{i\Gamma}$ of the configurations $(j+j', n+n')$ which are not geometric is negligible compared to (4.51), since $\mathcal{A}(j+j', n+n')$ decreases exponentially with large spins. The contribution of degenerate geometric configurations is also negligible, since one sums over a lower-dimensional sub-space in the space of spins.

Let $j$ and $\vec{n}$ be the background spin foam labels such that $(j, \vec{n})$ is a stationary point of $S_R(j, \vec{n})$ and all the 4-simplices have the Lorentzian geometry. Then we use the formulas from the $D = 3$ case, so that $e^{i\Gamma(j,\vec{n})/l_P^2}$ can be approximated by

$$N \sum_{j',\vec{n}'} e^{-\sum_f c_f \log(j_f + j_{f'}) + \frac{i\gamma}{l_P^2} S_R(j,\vec{n}) + \frac{\gamma}{2l_P^2} \langle S''_{R\,jj} j' j' + 2S''_{R\,jn} j' \vec{n}' + S''_{R\,nn} \vec{n}' \vec{n}' \rangle}$$

$$\approx N \sum_{j',\vec{n}'} e^{-\sum_f c_f [\log(j_f) + \frac{j'_f}{j_f}] + \frac{i\gamma}{l_P^2} S_R(j,\vec{n}) + \frac{\gamma}{2l_P^2} \langle \tilde{S}''_{R\,jj} j' j' + 2S''_{R\,jn} j' \vec{n}' + S''_{R\,nn} \vec{n}' \vec{n}' \rangle} \, ,$$

$$(4.52)$$

where $c_f = p_f m_f - 1$ and

$$\tilde{S}''_{R\,ff'} = S''_{R\,ff'} - i\frac{c_f}{j_f^2} \delta_{ff'} \approx S''_{R\,ff'} \, ,$$

for large $j_f$. The Gaussian sums in (4.52) can be approximated by the Gaussian integrals, so that we obtain

$$e^{i\Gamma(j,\vec{n})/l_P^2} \approx$$

$$N' e^{-\sum_f c_f \log(j_f) + \frac{i\gamma}{l_P^2} S_R(j,\vec{n}) - \frac{\gamma}{2l_P^2} Tr \log \tilde{S}''_R(j,\vec{n}) + \frac{1}{2\gamma} \sum_{f,f'} \tilde{c}_f \tilde{c}_{f'} G_{R\,ff'}(j,\vec{n})},$$

$$(4.53)$$

where

$$\tilde{S}''_R = \begin{pmatrix} S''_{R\,jj} & S''_{R\,jn} \\ S''_{R\,jn} & S''_{R\,nn} \end{pmatrix},$$

$G_{R\,ff'}$ is an element of the $jj$ block of the matrix $(S''_R)^{-1}$ and $\tilde{c}_f = c_f/j_f$. The equation (4.53) implies

$$\Gamma(j,\vec{n}) \approx \gamma S_R(j,\vec{n}) + l_P^2 \left( \sum_f c_f \log j_f + \frac{\gamma}{2} Tr \log \tilde{S}''_R(j,\vec{n}) \right)$$

$$+ l_P^2 \sum_{f,f'} c_f c_{f'} \frac{G_{R\,ff'}(j,\vec{n})}{2\gamma j_f j_{f'}},$$

$$(4.54)$$

where we have used $\Gamma \to Re\,\Gamma + Im\,\Gamma$ and we have omitted the constant $\log N'$. As in the $D = 3$ case, the dominant term is $S_R$, which is of $O(j)$, while the other terms are of subleading orders, namely of $O(\log j)$, $O(1/j)$ and $O(1/j^2)$, respectively.

The equation (4.54) implies that the classical limit of the EPRL/FK effective action for a Regge background $(j, \vec{n})$ is proportional to the Regge action

$$\gamma S_R(j,\vec{n}) = \gamma l_P^2 \sum_\Delta j_\Delta \delta_\Delta(\vec{n}) \approx \sum_\Lambda A_\Delta(L) \delta_\Delta(L),$$

$$(4.55)$$

if the triangle area satisfies $A_\Delta \approx \gamma l_P^2 j_\Delta$ for large $j$.

However, for an arbitrary $(j, \vec{n})$ background, it is not clear what is the classical action $\Gamma_0(j, \vec{n})$. This must be some generalization of the Regge action, which reduces to the Regge action when the configuration $(j, \vec{n})$ corresponds to an assignment of the edge lengths. In the limit of large spins and a smooth spacetime this would give some kind of a non-metric geometry. There are arguments that this non-metric geometry is the so-called twisted geometry, see [52, 118]. One can also argue that another candidate for this action is the area-Regge action [89], since it also defines a non-metric geometry which reduces to the metric one when the triangle areas correspond to the edge lengths via the Heron formula.

## 4.4 EA for spin-cube models

The problem with the classical limit of the EPRL/FK spin-foam model comes from the fact that the edge lengths are not the basic DoF. In the case of spin-cube models, see section 3.5, the basic DoF include the edge lengths, beside the labels for the triangles. Furthermore, the GR constraints can be imposed such that the triangle labels become the areas of the triangles corresponding to the edge lengths via the Heron formula, so that the independent variables are just the edge lengths.

The effective action $\Gamma(L)$ can be found as a solution of the EA equation, which takes the following form

$$e^{i\tilde{\Gamma}(L)} = \int_{D(L)} \prod_{\epsilon=1}^{E} \mu(L_\epsilon + l_\epsilon) \, dl_\epsilon \, \exp\left(i\tilde{S}_R(L+l) - i\sum_{\epsilon=1}^{E} \frac{\partial\tilde{\Gamma}}{\partial L_\epsilon} l_\epsilon\right), \quad (4.56)$$

where $\tilde{\Gamma} = \Gamma/l_P^2$ and

$$\tilde{S}_R = \frac{[S]_R}{l_P^2} = \frac{1}{l_P^2}\sum_{\Delta}[A_D]\,\delta_\Delta = \sum_{\Delta} m_\Delta\,\delta_\Delta\,,$$

is a Regge action with integer-valued areas. The integration region satisfies

$$D(L) \subset [-L_1,\infty) \times \cdots \times [-L_E,\infty)\,,$$

because $L_\epsilon > 0$ and $L_\epsilon + l_\epsilon > 0$ for all $\epsilon \in T(M)$.

Since we will be interested in the semiclassical limit $L_\epsilon \to \infty$, $\epsilon = 1, 2, \ldots, E$, we can replace $D(L)$ with $\mathbb{R}^E$. Then the equation (4.56) can be solved perturbatively as

$$\Gamma(L) = \sum_{n\geqslant 0}(l_P^2)^n\,\Gamma_n(L) + \text{const}\,, \quad (4.57)$$

where

$$\Gamma_0(L) = [S]_R(L) - il_P^2\sum_{\epsilon=1}^{E}\log\mu(L_\epsilon)\,,$$

see the Appendix C. Also, there will be restrictions for the large-$L$ asymptotics of the PI measure $\mu$, see section 4.5.

Note that the expansion (4.57) implies the following asymptotics for $L = (L_1, L_2, \ldots, L_E) \to (\infty, \ldots, \infty)$

$$\Gamma_n(L) = O(L^{2-2n})\,, \quad (4.58)$$

where the notation

$$f(L) = O(L^k)\,, \quad k \in \mathbb{Z}\,,$$

stands for

$$f(\lambda L) = O(\lambda^k),$$

when $\lambda \to \infty$.

The explicit form of the perturbative terms $\Gamma_n(L)$ can be obtained by introducing a formal perturbative parameter $\varepsilon \sim l_P^2$ such that

$$\Gamma(L, \varepsilon) = \sum_{n \geqslant 0} \varepsilon^n \Gamma_n(L) + \text{const}$$

where $\Gamma(L, \varepsilon)$ is a solution of

$$e^{i\Gamma/\varepsilon} = \int_{\mathbb{R}^E} \prod_{\epsilon=1}^{E} dl_\epsilon \, \exp\left(\frac{i}{\varepsilon} S_\mu(L+l) - \frac{i}{\varepsilon} \sum_{\epsilon=1}^{E} \frac{\partial \Gamma}{\partial L_\epsilon} l_\epsilon\right). \tag{4.59}$$

Here

$$S_\mu(L) = [S]_R(L) - i l_P^2 \sum_{\epsilon=1}^{E} \log \mu(L_\epsilon)$$

and the initial condition is $\Gamma_0 = S_\mu$.

By substituting the Taylor expansions for $[S]_R(L+l)$ and $\log \mu(L+l)$ into (4.59), one obtains

$$\Gamma_1(L) = \frac{i}{2} Tr\left(\log \hat{S}''_\mu(L)\right), \tag{4.60}$$

where

$$(\hat{S}''_\mu)_{\epsilon\epsilon'} = ([S]''_R)_{\epsilon\epsilon'} - ip \frac{\delta_{\epsilon,\epsilon'}}{L_\epsilon^2},$$

and we have taken that $\mu(L) = O(L^{-p})$ for large $L$ where $p > 1$, in order to ensure the finiteness of the path integral, see section 3.5.

A perturbative solution of (4.56) of the type (4.57) exists because the coefficients in the Taylor expansion

$$[S]_R(L+l) = [S]_R(L) + \sum_{\epsilon}([S]_R)'_\epsilon l_\epsilon + \frac{1}{2}\sum_{\epsilon,\epsilon'}([S]_R)''_{\epsilon,\epsilon'} l_\epsilon l_{\epsilon'} + \cdots,$$

satisfy

$$[S]_R^{(n)}(L) = O(L^{2-n}), \tag{4.61}$$

due to the fact that

$$[S]_R(L) = S_R(L) + l_P^2 \, \delta S_R(L),$$

where
$$\delta S_R = -\sum_{\Delta=1}^{F} \{A_\Delta(L)/l_P^2\}\,\theta_\Delta(L)\,,$$
and $\{x\} = x - [x]$ is the decimal part of a real number $x$.

The asymptotics (4.61) follows from the fact that $S_R(L)$ is a homogeneous function of degree 2 and $\delta S_R(L)$ is a homogeneous function of degree zero, while a partial derivative of a homogeneous function is a homogeneous function of the degree smaller by one.

Since $S_R(L) = O(L^2)$ and $\log \mu(L) = O(\log L)$, the terms in the expansion (4.57) satisfy
$$|\Gamma_n(L)| \gg |\Gamma_{n+1}(L)|\,,$$
for $n \geqslant 0$ and $L \to (\infty, \ldots, \infty)$, as well as
$$|S_R(L)| \gg |\sum_{\epsilon=1}^{E} \log \mu(L_\epsilon)| \gg |\delta S_R(L)|\,.$$
This implies that the classical limit of $\Gamma$ is the Regge action $S_R$, i.e.
$$\Gamma(L) \approx S_R(L)$$
for $L \to (\infty, \ldots, \infty)$. More precisely, when $L_\epsilon \gg l_P$ for all $\epsilon \in T(M)$.

Note that the solution (4.57) is not a real function, while a physical $\Gamma(L)$ has to be a real function. As we discussed in section 4.2, a real effective action can be obtained by the following transformation
$$\Gamma \to Re\,\Gamma + Im\,\Gamma\,. \tag{4.62}$$
The prescription (4.62) then gives a real effective action
$$\Gamma(L) \approx S_R(L) + l_P^2 \left( -\sum_{\epsilon=1}^{E} p\ln\left(\frac{L_\epsilon}{l_0}\right) + \delta S_R(L) + \frac{1}{2} Tr\left(\log S_R''(L)\right) \right), \tag{4.63}$$
when $L_\epsilon \gg l_P$ for all $\epsilon \in T(M)$. Here we have chosen the measure (3.77) in order to have a finite state-sum.

One can ask what is the smooth spacetime limit of the action (4.63)? We know that $S_R(L)$ gives the EH action, while the trace-log term gives the cut-off regularization of the field theory trace-log term, see section 4.6. However, the remaining terms do not have a smooth-manifold limit. In section 4.6 we will show that replacing the measure (3.77) with
$$\tilde{\mu} = \prod_{\epsilon=1}^{E} \left(1 + \frac{L_\epsilon^2}{l_0^2}\right)^{p/2}$$
is sufficient to make $\log \tilde{\mu} \to 0$ in the smooth limit. However, it is not clear what happens to $\delta S_R(L)$ in the smooth limit.

## 4.5 EA for the Regge model

As we saw in section 3.6, a Regge state-sum model is defined by fixing a spacetime triangulation $T(M)$ of a 4-manifold $M$ and by labeling the edges $\epsilon$ of $T(M)$ with non-negative numbers $L_\epsilon$, for space-like edges, or by $iL_\epsilon$, for time-like edges. The edge lengths will satisfy the triangle inequalities for every space-like triangle of $T(M)$. In the Lorentzian case, it is possible to have a triangulation where all the triangles are space-like, but as we discussed in section 3.6, it is more appropriate to choose some of the edge lengths to be time-like. The choice of the space-like and the time-like edge lengths will be reflected in the Regge action, which is defined according to the formula (2.26).

In order to completely specify a Regge state-sum model, we have to specify the set of values of $L_\epsilon$. The simplest and the most natural choice is the interval $(0, \infty)$. However, one can also consider $[a, \infty)$ and $[a, b]$ intervals, where $a, b > 0$. In addition, one can also consider the case of a discrete set of values, and we will consider the case $L_\epsilon = l_0 n$, $n \in \mathbb{N}$ and $l_0 > 0$.

Let $L_\epsilon \in (0, \infty)$. The corresponding state sum can be defined by the Regge path integral (3.101), which we write as

$$Z_R = \int_{D_E} d^E L \, \mu(L) \exp\left(iS_R(L)/l_P^2\right) , \qquad (4.64)$$

where $d^E L = \prod_{\epsilon=1}^{E} dL_\epsilon$. We use in (4.64) the Planck length instead of $\hbar$ because the EH action $S_{EH}$ is given by $S_R/G_N$, so that $S_{EH}/\hbar = S_R/l_P^2$.

The measure $\mu$ has to be chosen such that $Z_R$ is convergent and that the corresponding effective action allows a semiclassical expansion around the classical limit given by $S_R$. We will first study this problem for the measures which behave for large $L_\epsilon$ as

$$\mu(L) \approx \prod_{\epsilon=1}^{E} (L_\epsilon/L_1)^p \, e^{-(L_\epsilon/L_0)^\alpha} , \qquad (4.65)$$

where $L_0, p$ and $\alpha$ are parameters to be determined. $L_1$ is an arbitrary length, introduced only to make $\mu$ dimensionless. This class of measures is a simplified version of the exponentially damped 4-volume measures given by (3.109), and (4.65) will serve as a guide for analyzing the exponentially damped 4-volume case. We will restrict $\alpha$ to be non-negative number, so that when $\alpha > 0$ we will have a finite $Z$. When $\alpha = 0$, we will assume that $p$ is such that $Z$ is finite, see [87].

### 4.5.1　*One-dimensional toy model*

The effective action $\Gamma(L)$ can be defined by using the QFT effective action equation (4.2). However, in the QG case the "field variable" $L_\epsilon$ does not take all the values from $\mathbb{R}$, but $L_\epsilon > 0$. In order to see the difference it is sufficient to consider the case of a single variable $L \in [a, b]$. Let us start from the generating functional

$$Z(J) = \int_a^b dL\, \mu(L) \exp\left(\frac{i}{\hbar}S(L) + iJL\right) = e^{\frac{i}{\hbar}W(J)},$$

where $S(L)$ is a $C^\infty$ function. We define the "classical field", also known as the background field, as

$$\bar{L} = \frac{dW(J)}{dJ},$$

where $W(J) = -i\hbar \log Z(J)$.

The corresponding Legendre transformation is given by

$$\Gamma(\bar{L}) = W(J) - J\bar{L},$$

so that we obtain

$$e^{i\Gamma(\bar{L})/\hbar} = \int_{a-\bar{L}}^{b-\bar{L}} dl\, \mu(\bar{L}+l) \exp\left(\frac{i}{\hbar}[S(\bar{L}+l) - \Gamma'(\bar{L})\,l]\right), \qquad (4.66)$$

where $l = L - \bar{L}$ is the "quantum fluctuation". In the QFT case $a = -\infty$ and $b = \infty$, which gives the equation

$$e^{i\Gamma(L)/\hbar} = \int_{-\infty}^{\infty} dl\, \mu(L+l) \exp\left(\frac{i}{\hbar}[S(L+l) - \Gamma'(L)\,l]\right). \qquad (4.67)$$

In the simplest QG case we have $a = 0$ and $b = \infty$, so that

$$e^{i\Gamma(L)/\hbar} = \int_{-L}^{\infty} dl\, \mu(L+l) \exp\left(\frac{i}{\hbar}[S(L+l) - \Gamma'(L)\,l]\right). \qquad (4.68)$$

Let us look for a perturbative solution of (4.68) in the form

$$\Gamma(L) = S(L) + \hbar\Gamma_1(L) + \hbar^2\Gamma_2(L) + \cdots \qquad (4.69)$$

when $L \to \infty$ and with a measure which satisfies (4.65). The appearance of a semi-infinite interval of integration in the QG case may change the nature of the perturbative solution. Namely, the QFT perturbative expansion is based on the Gaussian integration formula

$$\int_{-\infty}^{\infty} e^{-zx^2/\hbar - wx}\, dx = \sqrt{\frac{\pi\hbar}{s}}\, e^{\hbar w^2/4z} = \sqrt{\pi\hbar}\, e^{-\frac{1}{2}\log z + \hbar w^2/4z}, \qquad (4.70)$$

where $Re\, z > 0$. Since in QFT $z = -iS''(\phi)/2$, then $Re\, z = 0$, but this problem is resolved by using the Euclidean metric, so that $z = S''_E(\phi) > 0$. In the QG case the integral (4.70) changes to

$$\int_{-L}^{\infty} e^{-zx^2/\hbar - wx}\, dx = \sqrt{\frac{\pi\hbar}{z}}\, e^{\hbar w^2/4z} \left[ \frac{1}{2} + \frac{1}{2}\,\mathrm{erf}\left( L\sqrt{\frac{s}{\hbar}} + \frac{\sqrt{\hbar}w}{2\sqrt{z}} \right) \right]$$

$$= \sqrt{\pi\hbar}\exp\left[ -\frac{1}{2}\log z + \frac{\hbar w^2}{4z} + \frac{\sqrt{\hbar}e^{-z\bar{L}^2/\hbar}}{2\sqrt{\pi z}\bar{L}}\left( 1 + O\left( \frac{\hbar}{z\bar{L}^2} \right) \right) \right], \quad (4.71)$$

where $\bar{L} = L + \hbar w/2z$ and

$$z/\hbar = -iS''/2l_P^2 - (\log\mu)''\,,$$

see Appendix C.

The key difference between (4.70) and (4.71) is that the argument of exp in (4.71) becomes a non-analytic function of $\hbar$. The non-analytic part is given by

$$\frac{\sqrt{\hbar}}{2\sqrt{\pi z}\,\bar{L}}\exp\left( -\frac{z\bar{L}^2}{\hbar} \right)\left[ 1 + O\left( \frac{\hbar}{z\bar{L}^2} \right) \right]\,, \quad (4.72)$$

and when (4.72) is expanded in powers of $\hbar$, it will contain both negative and positive powers of $\hbar$. The negative powers come from the $\exp(-z\bar{L}^2/\hbar)$ factor, which features an essential singularity at $\hbar = 0$. Then the EA equation cannot be solved as a series (4.69), since the perturbative expansion contains only the positive powers of $\hbar$. Such $\hbar$-series cannot be matched to the series with negative and half-integer powers of $\hbar$ introduced by (4.72), see Appendix C.

Hence the only way to solve the EA equation iteratively is to include negative and half-integer powers of $\hbar$ in the expansion (4.69), so that

$$\Gamma(L) = S(L) + \sum_{k\in\mathbb{Z}\backslash 0} \hbar^{k/2}\Gamma_k(L)\,. \quad (4.73)$$

However, the expansion (4.73) cannot be considered as a semiclassical expansion, because a semiclassical expansion is by definition an expansion in positive powers of $\hbar$. The reason for such a definition is that if the terms with negative powers of $\hbar$ are included, then the quantum corrections would dominate over the classical action when $\hbar \to 0$.

Therefore in order to have a semiclassical expansion for large $L$ we need to suppress the non-analytic term (4.72). This is an important requirement, because we know that for large length-scales the quantum gravity corrections are small. Hence we will require

$$\lim_{L\to\infty} \exp(-z\bar{L}^2/\hbar) = 0\,. \quad (4.74)$$

Since

$$\bar{L}^2 = L^2 + \hbar L w/z + \hbar^2 w^2/4z \approx L^2 \,,$$

for $L \to \infty$, the condition (4.74) is equivalent to

$$\operatorname{Re} z > 0 \,, \tag{4.75}$$

for $L \to \infty$. In the case when $\mu$ obeys the condition (4.65), we obtain

$$z/\hbar \approx -iS''(L)/2l_P^2 + p/L^2 + \alpha(\alpha - 1)(L_0)^{-\alpha} L^{\alpha-2} \,. \tag{4.76}$$

where we have replaced $\hbar$ with $l_P^2$. We also have

$$w = -i(\Gamma'_1 + l_P^2 \Gamma'_2 + \cdots) \,. \tag{4.77}$$

Therefore the condition (4.75) is equivalent to

$$p/L^2 + \alpha(\alpha - 1)L_0^{-\alpha} L^{\alpha-2} > 0 \,, \tag{4.78}$$

for $L \to \infty$. For $\alpha = 0$ the condition (4.78) gives $p > 0$. For $0 < \alpha < 1$ and any $p$, the condition (4.78) is not satisfied, so that there are no perturbative solutions. For $\alpha \geqslant 1$, the perturbative solutions are possible for any $p$.

### 4.5.2 *Higher-dimensional case*

Let us now try to generalize the analysis of the previous section to the case $E > 1$. Let $L = (L_1, \ldots, L_E) \in D_E$, then we obtain the following integro-differential equation

$$e^{i\Gamma(L)/l_P^2} = \int_{D_E(L)} d^E l \, \mu(L + l) \exp\left(iS_R(L + l)/l_P^2 - i \sum_\epsilon \frac{\partial \Gamma}{\partial L_\epsilon} l_\epsilon / l_P^2\right), \tag{4.79}$$

where the integration region $D_E(L)$ is a subset of $\mathbb{R}^E$ obtained by translating $D_E$ by a vector $-L$.

The main problem with generalizing the $E = 1$ results is that we do not know how to calculate exactly the integral

$$I_0 = \int_{D_E(L)} d^E l \exp\left(-\langle l, zl\rangle/l_P^2 + \langle w, l\rangle\right) \,,$$

where $z$ is an $E \times E$ symmetric complex matrix and $\langle x, y\rangle = \sum_{k=1}^E x_k y_k$. A reasonable conjecture is that for large $L$

$$I_0 \approx \int_{C_E(L)} d^E l \exp\left(-\langle l, zl\rangle/l_P^2 + \langle w, l\rangle\right) \,,$$

where

$$C_E(L) = [-L_1, \infty) \times \cdots \times [-L_E, \infty) \,.$$

From this conjecture it follows that

$$I_0 \approx \left(\frac{\pi l_P^2}{4}\right)^{E/2} (\det z)^{-1/2} \, e^{l_P^2 \langle w, z^{-1} w \rangle / 4} \prod_k \left[1 + \mathrm{erf}\left(\frac{\tilde{L}_k \sqrt{\lambda_k}}{l_P} + \frac{l_P \tilde{w}_k}{2\sqrt{\lambda_k}}\right)\right],$$

where $\lambda_k$ are the eigenvalues of the matrix $z$ and $\tilde{w} = Uw$, where $U$ is the matrix which puts $z$ in the diagonal form, i.e. $z = U^{-1} \mathrm{diag}(\lambda_1, \ldots, \lambda_E) \, U$.

By using the results of the $E = 1$ case, we conclude that a semiclassical solution of (4.79) will exist for the measures satisfying (4.65) when $L \gg l_P$, $\alpha \geqslant 1$ and any value of $p$. Note that a perturbative solution is also possible for $\alpha = 0$ and $p > 0$, but the convergence of the path integral is not guaranteed in that case.

Let us assume that $\alpha$ and $p$ are such that the perturbative expansion

$$\Gamma(L) = S_R(L) + l_P^2 \Gamma_1(L) + l_P^4 \Gamma_2(L) + \cdots \tag{4.80}$$

is valid. We will try to derive some restrictions on $\alpha$ and $p$ from the requirement that

$$\frac{l_P^2 |\Gamma_{n+1}(L)|}{|\Gamma_n(L)|} \ll 1 \tag{4.81}$$

for $L/l_P \gg 1$ and for all $n$.

The requirement (4.81) defines the semiclassical expansion, since it implies that the quantum corrections are much smaller than the classical value. A weaker version of (4.81) is

$$\frac{l_P^2 |\Gamma_{n+1}(L)|}{|\Gamma_n(L)|} < 1 \tag{4.82}$$

for $L/l_P > 1$ and all $n$, and in this case we will consider the solution (4.80) to be perturbative. Note that the series (4.80) is an asymptotic series, so that it does not have to converge, and therefore (4.82) is a way to estimate the region where the perturbative expansion is valid.

Let $\alpha > 0$ and if the perturbative expansion (4.80) is valid we can use the approximation $D_E(L) \approx \mathbb{R}^E$ to solve the equation (4.79). We obtain

$$\Gamma_1(L) = i \sum_{\epsilon=1}^{E} [(L_\epsilon/L_0)^\alpha - p \ln(L_\epsilon/L_1)] + \frac{i}{2} Tr \log S_R''(L), \tag{4.83}$$

which is of $O(L^\alpha)$.[3] The higher-order corrections $\Gamma_n$ can be determined by using the diagrammatic techniques from QFT, see [69]. These corrections

---

[3]We define $f(x_1, x_2, \ldots, x_n) = O(x^\alpha)$ if $f(\lambda x_1, \lambda x_2, \ldots, \lambda x_n) \approx \lambda^\alpha g(x_1, x_2, \ldots, x_n)$ for $\lambda \to \infty$.

can be evaluated by using the effective action diagrams (EAD), whose $k$-valent vertices ($k \geqslant 3$) carry the weights $S_k = i S_R^{(k)}(L)/k!$ and the edges carry the propagator $G(L) = i(S_R'')^{-1}$. The contributions from a non-trivial path-integral measure can be taken into account if in the formulas for the vertex weights and the propagator we replace $S_R$ by

$$\bar{S}_R = S_R + i l_P^2 \sum_{\epsilon=1}^{E} [(L_\epsilon/L_0)^\alpha - p \ln(L_\epsilon/L_1)] .$$

This follows from the EA equation (4.79), since it can be rewritten as the EA equation with a trivial measure term and the action $\bar{S}$. The perturbative solution will take the form

$$\Gamma = \bar{S}_R + l_P^2 \bar{\Gamma}_1 + l_P^4 \bar{\Gamma}_2 + \cdots ,$$

where $\bar{\Gamma}_n$ will be given by the EAD with $\bar{G}$ propagator and $\bar{S}_k$ vertices. Since

$$\bar{\Gamma}_n = \Gamma_{n,0} + l_P^2 \bar{\Gamma}_{n,1} + l_P^4 \bar{\Gamma}_{n,2} + \cdots ,$$

we obtain

$$\Gamma = S_R + l_P^2(-i \log \mu + \Gamma_{1,0}) + l_P^4(\Gamma_{2,0} + \bar{\Gamma}_{1,1}) + l_P^6(\Gamma_{3,0} + \bar{\Gamma}_{1,2} + \bar{\Gamma}_{2,1}) + \cdots .$$
$$(4.84)$$

For example,

$$\Gamma_2 = \langle (S_3)^2 G^3 \rangle + \langle S_4 G^2 \rangle + Res \left[ l_P^{-4} \, Tr \log \bar{G} \right] ,$$

$$\Gamma_3 = \langle (S_3)^4 G^6 \rangle + \langle S_3 S_4 G^4 \rangle + \langle S_6 G^3 \rangle$$

$$+ Res \left[ l_P^{-6} \left( Tr \log \bar{G} + \langle (\bar{S}_3)^2 \bar{G}^3 \rangle + \langle \bar{S}_4 \bar{G}^2 \rangle \right) \right] ,$$

and so on. Here $\langle XY \cdots \rangle$ denotes the sum of all possible contractions of the tensors $X, Y, \ldots$, which is given by the corresponding EAD, see [69]. The residuum terms are determined by the formula

$$Res \, (z^{-n} f(z)) = \frac{f^{(n-1)}(0)}{(n-1)!} ,$$

where $z = l_P^2$.

By using that $S_n = O(L^{2-n})$ and $\bar{S}_n = O(L^{2-n}) + O(L^{\alpha-n})$, we obtain

$$\Gamma_{n+1,0} = O(L^{-2n}), \quad \bar{\Gamma}_{n+1-k,k} = O(L^{k\alpha-2n}),$$

where $k = 1, 2, \ldots, n$. Consequently

$$l_P^{2n} \Gamma_{n+1,0} = O((l_P/L)^{2n}), \quad l_P^{2n} \bar{\Gamma}_{n+1-k,k} = O((L/L_0)^{k\alpha}(l_P/L)^{2n}).$$

Hence

$$l_P^{2n} \Gamma_{n+1}(L) = O\left((l_P/L)^{2n}\right) + O\left((l_P/L)^{2n}(L/L_0)^{n\alpha}\right)$$

$$= O\left((l_P/L)^{2n}\right) + O\left((L/L_s)^{n(\alpha-2)}\right),$$

where

$$L_s = \left(L_0^\alpha / l_P^2\right)^{\frac{1}{\alpha-2}} \tag{4.85}$$

is a new length scale which together with $l_P$ will determine the validity of the semiclassical expansion.

By using the criterion (4.81) we obtain that the perturbative expansion (4.84) will be semiclassical if

$$l_P^2/L^2 \ll 1, \quad (L/L_s)^{\alpha-2} \ll 1. \tag{4.86}$$

The condition (4.86) will be satisfied if $L \gg l_P$ and $L \gg L_s$ for $\alpha < 2$, while for $\alpha > 2$ we need that $l_P \ll L \ll L_s$.

When $\alpha = 2$, we have

$$l_P^{2n} \Gamma_{n+1}(L) = O\left((l_P/L)^{2n}\right) + O\left((l_P/L_0)^{2n}\right),$$

so that the series (4.84) will be semiclassical for $L_0 \gg l_P$ and $L \gg l_P$.

Note that the $\Gamma_1$ takes imaginary number values, and the higher-order quantum corrections $\Gamma_n$ will in general take complex number values. The same happens in QFT, and since we want to have a real effective action, we have to restrict our complex solution to real values. As we discussed in section 4.2, this problem can be solved by

$$\Gamma(L) \to Re\,\Gamma(L) \pm Im\,\Gamma(L). \tag{4.87}$$

Since the transformation (4.87) is metric-independent, it can be used in QG to define a real effective action, while the sign ambiguity can be fixed by an experimental input.

### 4.5.3 *Cosmological constant measures*

When $\alpha > 2$ the expansion (4.84) will be semiclassical if $l_P \ll L \ll L_s$, so that we need that $L_s \gg l_P$, which is satisfied if $L_0 \gg l_P$. The interesting case is $\alpha = 4$, because the quantum Regge calculus measures (3.109) are of this type. More generally, one can consider the measures which satisfy

$$\log \mu = O(L^4), \tag{4.88}$$

and $\log \mu < 0$ for $L$ large.

Let us consider the following PI measure

$$\mu_c(L) = \exp\left(-\sum_{\sigma=1}^{N} V_\sigma(L)/L_0^4\right) = \exp\left(-V_4(L)/L_0^4\right). \qquad (4.89)$$

This measure is a special case of the measure (3.109), where we have taken $p = 0$ for the sake of simplicity. Given that $V_\sigma(L) = O(L^4)$ for large $L$, we have

$$\log \mu_c(L) = O((L/L_0)^4),$$

so that we can apply the same reasoning when calculating the perturbative effective action as in the case of the $\alpha = 4$ measure (4.65) with $p = 0$.

Namely, if we substitute $D_E(L)$ in (4.79) by $\mathbb{R}^E$ we obtain

$$\Gamma_1(L) = i\frac{V_4(L)}{L_0^4} + \frac{i}{2} Tr \log S_R''(L), \qquad (4.90)$$

so that $\Gamma_1(L) = O((L/L_0)^4)$ and

$$\bar{S}_R = S_R + i\, l_P^2 V(L)/L_0^4.$$

Hence the formulas from the previous section give

$$\Gamma = \bar{S}_R + l_P^2 \bar{\Gamma}_1 + l_P^4 \bar{\Gamma}_2 + \cdots = S_R + l_P^2 \Gamma_1 + l_P^4 \Gamma_2 + \cdots,$$

where

$$l_P^{2n} \Gamma_{n+1}(L) = O\left((l_P/L)^{2n}\right) + O\left((L/L_s)^{2n}\right),$$

and

$$L_s = \frac{L_0^2}{l_P}. \qquad (4.91)$$

Therefore the effective action will have a semiclassical expansion if

$$l_P \ll L_\epsilon \ll \frac{L_0^2}{l_P}, \qquad (4.92)$$

which is satisfied for $L_0 \gg l_P$.

If we define the physical effective action as

$$S_{eff} = \frac{1}{G_N} \left(Re\,\Gamma \pm Im\,\Gamma\right),$$

we then obtain from (4.90)

$$S_{eff} = \frac{S_R}{G_N} \pm \frac{l_P^2}{G_N L_0^4} V_4 \pm \frac{l_P^2}{2G_N} Tr \log S_R'' + O(l_P^4). \qquad (4.93)$$

Hence the second term in (4.93) can be interpreted as the cosmological constant term with the value of the cosmological constant given by

$$\Lambda = \pm \frac{l_P^2}{2\,L_0^4} = \pm \frac{1}{2L_s^2}\,. \tag{4.94}$$

Note that the value (4.94) will be very small in units of $l_P^{-2}$, since

$$l_P^2 |\Lambda| = \frac{1}{2} \left( \frac{l_P}{L_0} \right)^4$$

and (4.92) gives $l_P/L_0 \ll 1$. Therefore we have a mechanism to generate a small cosmological constant from the PI measure (4.89), as a first-order quantum correction. If we define $L_\Lambda = 1/\sqrt{\Lambda}$, then the observed value for $\Lambda$ gives $L_\Lambda \approx 10^{26}\,m$, or in dimensionless units, $l_P^2 \Lambda \approx 10^{-122}$. Since $L_s = L_\Lambda/\sqrt{2}$, we get $L_0 \approx 10^{-5}\,m \approx 10\,\mu m$, so that $l_P/L_0 \approx 10^{-30.5} \ll 1$.

Observe that $\Gamma_3(L) = O(L^4)$, so that one can have an $O(l_P^6/L_0^8)$ correction to the CC value (4.94). Hence the exact value of the cosmological constant will be given by

$$\Lambda = \pm \frac{l_P^2}{2\,L_0^4} \left( 1 + c_3\,\frac{l_P^4}{L_0^4} \right), \tag{4.95}$$

where $c_3$ is a numerical constant of $O(1)$. Since $l_P/L_0 \approx 10^{-30.5}$ for the observed value of the cosmological constant, this correction can be neglected.

In the case of the measure (3.109) with $p \neq 0$ we will obtain the same formulas for the cosmological constant. This is because the corresponding terms in the $\log \mu$ term are of subleading order in $L$ with respect to the $V_4(L)$ term.

### 4.5.4   *Regge EA for non-zero CC*

The path integral for a classical theory given by the Regge action plus the cosmological constant term, is given by

$$Z = \int_{D_E} \mu(L)\, d^E L\, \exp\left( iS_{Rc}(L)/l_P^2 \right), \tag{4.96}$$

where $D_E$ is a subset of $\mathbb{R}_+^E$ where the appropriate inequalities on $L_\epsilon$ hold (depending on the choice of the space-like and the time-like edges) and

$$S_{Rc} = \sum_{\Delta=1}^{F} A_\Delta(L)\delta_\Delta(L) + \Lambda_c\, V_4(L), \tag{4.97}$$

is the Regge action corresponding to the Einstein-Hilbert action with the CC term and $\Lambda_c$ is the classical value of the cosmological constant. $A_\Delta$ is the area of a triangle $\Delta$, $\theta_\Delta$ is the deficit angle and $V_4$ is the 4-volume of $T(M)$.

We will consider only a finite-volume $T(M)$, and $M = \Sigma \times [a, b]$, so that for a compact $\Sigma$ the corresponding $T(M)$ will be always of a finite 4-volume, while for a non-compact $\Sigma$, we will consider finite vectors $L$, so that non-zero $L_\epsilon$ are assigned to $B_3(\Sigma) \times [a, b]$, where $B_3$ is a ball in $\Sigma$.

We will also introduce a classical CC length scale $L_c$ such that

$$\Lambda_c = \pm \frac{1}{2L_c^2} , \qquad (4.98)$$

and we will choose the PI measure $\mu(L)$ as

$$\mu(L) = \exp\left(-V_4(L)/L_0^4\right) , \qquad (4.99)$$

where $L_0$ is a new length scale. This is the $p = 0$ case of the measure (3.109). Since the perturbative analysis does not depend on the value of $p$, we will take $p = 0$ for the sake of simplicity. We saw in the previous section that the measure (4.99) generates a small quantum correction to the classical CC when $\Lambda_c = 0$ and $L_0 \gg l_P$. This is also the simplest local measure which allows a perturbative effective action for large $L_\epsilon$ and which is manifestly diffeomorphism invariant in the smooth limit ($L_\epsilon \to 0$, $E \to \infty$), see section 4.6.

We must remember that the smooth limit is an approximation in our approach, since we are postulating that the spacetime triangulation is physical, so that $E$ is fixed and $L_\epsilon$ are non-zero. We will also assume that $E$ is a large number, i.e. $E \gg 1$, so that the physical piecewise linear manifold $T(M)$ looks like the smooth manifold $M$ for small $L_\epsilon$.

The quantum effective action $\Gamma(L)$ associated to the theory defined by the path integral (4.96) is determined by the following integro-differential equation

$$e^{i\Gamma(L)/l_P^2} = \int_{D_E(L)} \mu(L+l) \, d^E l \, \exp\left(iS_{Rc}(L+l)/l_P^2 - i\sum_{\epsilon=1}^{E} \frac{\partial \Gamma}{\partial L_\epsilon} l_\epsilon/l_P^2\right) ,$$

$$(4.100)$$

where $D_E(L)$ is a subset of $\mathbb{R}^E$ obtained by translating the region $D_E$ by the vector $-L$. When $L_\epsilon \to \infty$, $\epsilon = 1, 2, \ldots, E$, then $D_E(L) \to \mathbb{R}^E$, and we can assume that the perturbative solution of (4.100) will be well approximated by the perturbative solution of the equation

$$e^{i\Gamma(L)/l_P^2} = \int_{\mathbb{R}^E} d^E l \, \exp\left(i\bar{S}_{Rc}(L+l)/l_P^2 - i\sum_{\epsilon=1}^{E} \frac{\partial \Gamma}{\partial L_\epsilon} l_\epsilon/l_P^2\right) , \qquad (4.101)$$

where

$$\bar{S}_{Rc}(L) = S_{Rc}(L) + il_P^2 V_4(L)/L_0^4 \,, \tag{4.102}$$

see section 4.5.2. The perturbative solution of (4.101) can be written as

$$\Gamma = \bar{S} + l_P^2 \bar{\Gamma}_1 + l_P^4 \bar{\Gamma}_2 + \cdots \,, \tag{4.103}$$

where $\bar{\Gamma}_n$ will be given by the EA diagrams constructed for the action $\bar{S}_{Rc}$, see section 4.5.2. Since

$$\bar{\Gamma}_n = \Gamma_{n,0} + l_P^2 \bar{\Gamma}_{n,1} + l_P^4 \bar{\Gamma}_{n,2} + \cdots \,, \tag{4.104}$$

we obtain

$$\Gamma = S_{Rc} + l_P^2(-i\log\mu + \Gamma_{1,0}) + l_P^4(\Gamma_{2,0} + \bar{\Gamma}_{1,1}) + l_P^6(\Gamma_{3,0} + \bar{\Gamma}_{1,2} + \bar{\Gamma}_{2,1}) + \cdots \,. \tag{4.105}$$

Hence

$$\Gamma_n(L) = D_n(L) + R_n(L) \,, \tag{4.106}$$

where $D_n$ is the contribution from the $n$-loop EA diagrams for the action $S_{Rc}$, while

$$R_n = Res_n \sum_{k=1}^{n-1} \bar{D}_k \,, \tag{4.107}$$

where

$$Res_n \, f(l_P^2) = \lim_{l_P^2 \to 0} \frac{f^{(n)}(l_P^2)}{n!} \,. \tag{4.108}$$

The $\bar{D}_k$ terms are defined as

$$\bar{D}_n(L) = D_n(L, \bar{L}_c^2) \,, \tag{4.109}$$

where

$$\bar{L}_c^2 = L_c^2 \left(1 + il_P^2 L_c^2/L_0^4\right)^{-1} = L_c^2 \left(1 + il_P^2/L_{0c}^2\right)^{-1} \,. \tag{4.110}$$

In order for the measure contributions to be perturbative, we see from (4.110) that we need $l_P/L_{0c} < 1$, which is equivalent to

$$L_0 > \sqrt{l_P L_c} \,. \tag{4.111}$$

We will study the case $L_\epsilon > L_c$, since the perturbative analysis is simpler than in the $L_\epsilon < L_c$ case. The large-$L$ asymptotics of $\bar{\Gamma}_n(L)$ functions can be determined from

$$S_n(L) = O(L^{4-n})/L_c^2 \,, \tag{4.112}$$

and the formula for the EA diagrams, see (4.118). Consequently, for $n > 1$

$$D_n(L) = O\left(\left(L_c^2/L^4\right)^{n-1}\right), \tag{4.113}$$

where the $O$ notation is defined as

$$f(L) = O(L^a) \Leftrightarrow f(\lambda L) \approx \lambda^a g(L) \tag{4.114}$$

when $\lambda \to \infty$. Since

$$\bar{\Gamma}_n(L) = D_n(L, \bar{L}_c^2), \tag{4.115}$$

we obtain

$$\bar{\Gamma}_n(L) = O\left(\left(\bar{L}_c^2/L^4\right)^{n-1}\right). \tag{4.116}$$

The asymptotics (4.113) can be derived by considering the one-dimensional ($E = 1$) toy model

$$S_{Rc} = \left(L^2 + \frac{L^4}{L_c^2}\right)\theta(L), \tag{4.117}$$

where $\theta(L)$ is a homogeneous $C^\infty$ function of degree zero. Consequently

$$D_n(L) = \sum_{l \in \mathbb{N}} c_{nl}\,(G(L))^{k_l}\, S_{n_1}(L) \cdots S_{n_l}(L), \tag{4.118}$$

where $G = 1/S''_{Rc}$, $S_n = S_{Rc}^{(n)}/n!$, $k_l$ is the number of edges of an $n$-loop EA graph with $l$ vertices and $c_{nl}$ are numerical factors.

The asymptotics (4.113) implies that there are no $O(L^4)$ terms in $D_n(L)$, and hence $D_n(L)$ cannot contribute to the effective CC. This also happens for the $R_n$ terms, which can be seen from the toy model, where

$$\bar{S}''_{Rc} = \theta_1(L)[1 + (L^2/\bar{L}_c^2)\,\theta_2(L)], \tag{4.119}$$

and $\theta_k$ are homogeneous functions of degree zero. Consequently

$$\log \bar{S}''_{Rc} = \log(L^2/\bar{L}_c^2) + \log \theta_1(L) + \log\left[1 + O(\bar{L}_c^2/L^2)\right], \tag{4.120}$$

while from (4.116) it follows that

$$R_n(L) = O((L_{0c}^2)^{-n+1}) + O(L^{-2}(L_{0c}^2)^{-n+2}) + O(L^{-4}(L_{0c}^2)^{-n+3}) + \cdots. \tag{4.121}$$

We then obtain

$$\Gamma_1 = O(L^4/L_0^4) + \log O(L^2/L_c^2) + \log \theta_1(L) + O(L_c^2/L^2), \tag{4.122}$$

and

$$\Gamma_n = D_n + R_n = O((L_c^2/L^4)^{n-1}) + L_{0c}^{2-2n}\,O(L_c^2/L^2) = L_{0c}^{2-2n}\,O(L_c^2/L^2). \tag{4.123}$$

Note that we have discarded the constant pieces in $\Gamma_n(L)$.

Hence there are no $O(L^4)$ terms in $\Gamma_n$ for $n > 1$ and therefore the effective cosmological constant will be determined by the $\log \mu$ term, so that

$$\Lambda_g = \Lambda_c + \Lambda_\mu = \pm \frac{1}{2L_c^2} \pm \frac{l_P^2}{2L_0^4}. \tag{4.124}$$

The formula (4.124) follows from the physical effective action, which is defined as

$$S_{eff} = (Re\,\Gamma \pm Im\,\Gamma)/G_N. \tag{4.125}$$

We have used in (4.125) the QG Wick rotation

$$\Gamma \to Re\,\Gamma \pm Im\,\Gamma, \tag{4.126}$$

in order to make the effective action a real function, since the solutions of the EA equation are complex, see section 4.2. The sign ambiguity in (4.125) will be fixed by requiring that $\Lambda_\mu$ is positive, see section 5.3.

Note that the condition (4.111) and $L_\epsilon > L_c$ ensure that the effective action is semiclassical, which implies that the quantum corrections to the classical action will be small for

$$L_0 \gg \sqrt{l_P L_c}. \tag{4.127}$$

In this case

$$|S_{Rc}|/l_P^2 \gg |\Gamma_1| = \left|\log \mu - \frac{1}{2}Tr \log S_{Rc}''\right|, \tag{4.128}$$

and

$$|\Gamma_n| \gg l_P^2 |\Gamma_{n+1}|, \tag{4.129}$$

for all $n$.

Also note that the effective action will remain semiclassical if $L_c$ is large and $L_\epsilon < L_c$, but in this case we need $L_\epsilon \gg l_P$ in addition to the condition (4.111). This can be seen from the asymptotics of $\bar{\Gamma}_n(L)$ terms when $L_\epsilon < L_c$, since

$$\log \bar{S}''(L) = \log \theta_1(L) + \log \left[1 + O(L^2/\bar{L}_c^2)\right] \tag{4.130}$$

and

$$\bar{\Gamma}_{n+1}(L) = O(1/L^{2n}) \left[1 + O(L^2/\bar{L}_c^2)\right]. \tag{4.131}$$

The asymptotics (4.130) and (4.131) imply that we may have terms of $O(l_P^{2n})$ for any $n$ contributing to $\Lambda_g$. However, since $\Lambda_g$ is a constant, i.e. it is independent of $L$, and given that we showed for $L_\epsilon > L_c$ that there are no such terms in $\Lambda_g$, see equation (4.124), then this implies that in the case $L_\epsilon < L_c$ the sum of $O(l_P^{2n})$ terms must be zero. Hence one obtains the same formula for $\Lambda_g$ as (4.124).

We will consider an edge length $L_\epsilon$ to be large if $L_\epsilon \gg l_P$, so that a triangulation will have large edge lengths if

$$L_\epsilon \geqslant L_K \gg l_P, \tag{4.132}$$

where $L_K$ is the minimal edge length. The length $L_K$ will serve as a QFT cutoff in the smooth-manifold approximation of the effective action, see section 4.6.

### 4.5.5 Discrete-length Regge models

Let us now analyze the effective action for discrete-length Regge state-sum models. In this case we can write $L_\epsilon = \gamma n_\epsilon l_P$ where $n \in \mathbb{N}$ and $\gamma > 0$. Then

$$Z_R = \sum_{n \in N_{\gamma,E}} \mu(L(n)) \exp\left(i S_R(L(n))/l_P^2\right),$$

where $N_{\gamma,E}$ is a subset of $\mathbb{N}^E$ such that $\gamma n l_P \in D_E$.

The effective action equation is given by

$$e^{i\Gamma(L)/l_P^2} = \sum_{n \in N_{\gamma,E}} \mu(L+l(n)) \exp\left(i S_R(L + l(n))/l_P^2 - i \sum_\epsilon \frac{\partial \Gamma}{\partial L_\epsilon} l_\epsilon(n)/l_P^2\right),$$

$$\tag{4.133}$$

where $l_\epsilon(n) = \gamma l_P n_\epsilon - L_\epsilon$. Note that the variable $L$ in (4.133) is actually a quantum expectation value of $L$, which we have denoted as $\bar{L}$, see (4.66). We will refer to this $L$ as the background $L$, and it can take any value in $\mathbb{R}^E$. However, for the sake of simplicity we will consider the backgrounds such that $L/\gamma l_P \in \mathbb{Z}^E$, so that $l/\gamma l_P \in \mathbb{Z}^E$. Otherwise $l_\epsilon/\gamma l_P = m_\epsilon + x_\epsilon$ where $m_\epsilon \in \mathbb{Z}$ and $x_\epsilon \in (0,1)$.

One expects to obtain the same results for the semiclassical solution of (4.133) as in the continuous case. However, there is an obstruction, due to the fact that

$$\sum_{m=-k}^{\infty} f(m) \neq \int_{-k}^{\infty} f(l)\, dl.$$

In our case this problem appears when computing the one-loop correction, which is given by the logarithm of

$$\sum_{m \in \mathbb{Z}^E} \exp\left(\frac{i}{2}\langle \gamma m, S_R''(L)\gamma m\rangle\right).$$

Since

$$\sum_{m \in \mathbb{Z}} \exp(iam^2) \neq \sqrt{\frac{i\pi}{a}} \, ,$$

we cannot use the Gaussian integral approximation. However, one can show that

$$\sum_{m \in \mathbb{Z}} \exp(iam^2) \approx \sqrt{\frac{i\pi}{a}} \, ,$$

for $a \to 0$ (see Appendix D). Hence

$$\sum_{m \in \mathbb{Z}^E} \exp\left(\frac{i}{2}\langle m, \gamma^2 S_R''(L) \, m \rangle\right) \approx (2i\pi)^{E/2} \left(\det(\gamma^2 S_R''(L))^{-1/2} \, , \quad (4.134)$$

only if the entries of the Hessian matrix $S_R''(L)$ satisfy

$$\gamma^2 |S_R''(L)| \ll 1 \, . \tag{4.135}$$

Since $S_R''(L) = O(1)$, we need $\gamma^2 \ll 1$ which implies $\gamma \ll 1$.

Therefore the semiclassical approximation will be valid only if the spectrum gap is much smaller than $l_P$. This is a surprising result, since it implies that in the natural case when the spectrum gap is of order $l_P$, which corresponds to $\gamma \approx 1$, one cannot solve the EA equation (4.133) perturbatively. Even if we abandon the positivity of $L + l$, and replace $D_E(L)$ with $\mathbb{R}^E$, the result (4.135) holds.

The requirement (4.135) can be also applied to the semiclassical approximation of the effective action for spin-foam models. In the spin foam case, instead of the edge lengths, we have the triangle area variables $j \, l_P^2$, such that $j_f \in \mathbb{N}/2$, $f = 1, 2, \ldots, F$ and $S_R(L)/l_P^2 \to S(j)$ where

$$S(j) \approx \sum_{f=1}^{F} j_f \, \theta_f(j) \, , \tag{4.136}$$

for $j_f \gg 1$ and $\theta(j) = O(1)$, see section 4.3.2. Hence $\gamma = 1/2$ in the spin foam case. However, there is no problem for the semiclassical approximation, since the Hessian satisfies $S''(j) = O(1/j)$. Therefore $|S''(j)| \ll 1$ so that

$$\sum_{m \in \mathbb{Z}^F} \exp\left(\frac{i}{8}\langle m, S''(j) \, m \rangle\right) \approx (8i\pi)^{F/2} \left(\det(S''(j))^{-1/2} \, .$$

Note that the spin foam analog of the $D_E(L)$ integration region is

$$D_F(j) = [-j_1, \infty) \times \cdots \times [-j_F, \infty) \, .$$

In order to have a perturbative expansion of the SF effective action for large $j_f$ we need to modify the standard SF measure

$$\mu(j) = \prod_{f=1}^{F} \dim j_f = \prod_{f=1}^{F} (2j_f + 1),$$

by including an exponentially damping term. For example, the measure

$$\tilde{\mu}(j) = \prod_{f=1}^{F} (2j_f + 1) \, e^{-c(j_f)^\alpha},$$

where $c > 0$ and $1 > \alpha > 0$, will allow a semiclassical solution. The parameter $\alpha$ must be less than 1 in order for (4.136) to be dominant over

$$\Delta \Gamma_1 = c \sum_{f=1}^{F} (j_f)^\alpha \tag{4.137}$$

for large $j_f$, since (4.137) is the contribution of the new term in the measure to the effective action.

In order to have an explicit LQG interpretation of the modified measure, one can replace $(j_f)^\alpha$ with $(j_f(j_f + 1))^{\alpha/2}$, where $j_f(j_f + 1)$ is an $SU(2)$ Casimir operator eigenvalue.

### 4.5.6  *Regge EA for gravity and matter*

In order to see what is the effect of matter on the value of the cosmological constant, we will consider a scalar field $\phi$ on a 4-manifold $M$ with a metric $g$ such that the scalar-field action is given by

$$S_s(g, \phi) = \frac{1}{2} \int_M d^4x \sqrt{|g|} \left[ g^{\mu\nu} \, \partial_\mu \phi \, \partial_\nu \phi - U(\phi) \right], \tag{4.138}$$

where $U(\phi)$ is a polynomial of the degree greater or equal than 2.

When the metric $g$ is non-dynamical, the EoM of (4.138) are invariant under the constant shifts of the potential $U$. However, we know that the metric is dynamical, so that the constant shifts in $U$ will give contributions to the cosmological constant term. These classical shifts of the potential will affect the value of $\Lambda_c$, so we will therefore assume that $\Lambda_c \neq 0$.

On $T(M)$ the action (4.138) becomes

$$S_{Rs} = \frac{1}{2} \sum_\sigma V_\sigma(L) \sum_{k,l} g_\sigma^{kl}(L) \, \phi_k' \, \phi_l' - \frac{1}{2} \sum_\pi V_\pi^*(L) \, U(\phi_\pi), \tag{4.139}$$

where $g_\sigma^{kl}$ is the inverse matrix of the metric in a 4-simplex $\sigma$

$$g_{kl}^{(\sigma)} = \frac{L_{0k}^2 + L_{0l}^2 - L_{kl}^2}{2L_{0k} L_{0l}}, \tag{4.140}$$

while $\phi'_k = (\phi_{\pi_k} - \phi_{\pi_0})/L_{0k}$ and $V_\pi^*$ is the volume of the dual cell for a vertex point $\pi$ of $T(M)$, see section 2.2.

The quantum corrections due to gravity and matter fluctuations can be described by the effective action based on the classical action

$$S(L, \phi) = \frac{1}{G_N} S_{Rc}(L) + S_{Rs}(L, \phi), \qquad (4.141)$$

where $S_{Rc}$ is the Regge action with the CC term. Since

$$S(L, \phi)/\hbar = S_{Rc}(L)/l_P^2 + G_N S_{Rs}(L, \phi)/l_P^2 \equiv S_{Rm}(L, \phi)/l_P^2, \qquad (4.142)$$

the EA equation becomes

$$e^{i\Gamma(L,\phi)/l_P^2} = \int_{D_E(L)} d^E l \int_{\mathbb{R}^V} \prod_\pi d\chi_\pi \exp\left[ i\bar{S}_{Rm}(L + l, \phi + \chi)/l_P^2 \right.$$
$$\left. - i\sum_\epsilon \frac{\partial\Gamma}{\partial L_\epsilon} l_\epsilon/l_P^2 - i\sum_\pi \frac{\partial\Gamma}{\partial\phi_\pi} \chi_\pi/l_P^2 \right], \qquad (4.143)$$

where $\bar{S}_{Rm} = \bar{S}_{Rc} + G_N S_{Rs}(L, \phi)$ and $\bar{S}_{Rc} = S_{Rc} - il_P^2 \log\mu$.

Since we are using an exponentially damped PI measure for the $L$ variables, we can use the approximation $D_E(L) \approx \mathbb{R}^E$ when $L_\epsilon \to \infty$, see section 4.5.2. We can then solve (4.143) perturbatively in $l_P^2$ by using the EA diagrams for the action $\bar{S}_{Rm}$.

It is convenient to introduce a dimensionless field $\sqrt{G_N}\,\phi$, so that one has $\sqrt{G_N}\,\phi \to \phi$ and $S_{Rm} = S_{Rc} + S_{Rs}$. The perturbative solution will be given by

$$\Gamma(L, \phi) = S_{Rm}(L, \phi) + l_P^2 \Gamma_1(L, \phi) + l_P^4 \Gamma_2(L, \phi) + \cdots, \qquad (4.144)$$

where $\Gamma_n$ are given by the EA diagrams corrected by the measure contributions, see section 4.5.2.

It is not difficult to see that

$$\Gamma(L, \phi) = \Gamma_g(L) + \Gamma_m(L, \phi), \qquad (4.145)$$

and that for constant $\phi$ configurations

$$\Gamma_m(L, \phi) = V_4(L)\, U_{eff}(\phi). \qquad (4.146)$$

We expect that the expansion (4.144) will be semiclassical for $L \gg l_P$ and $\phi \ll 1$. This can be verified by studying the one-dimensional ($E = 1$) toy model for the potential

$$U(\phi) = \frac{\omega^2}{2}\phi^2 + \frac{\lambda}{4!}\phi^4, \qquad (4.147)$$

where $\hbar\omega = m$ is the matter field mass and $\lambda$ is the matter self-interaction coupling constant. The toy-model classical action can be taken to be

$$S_{Rm}(L,\phi) = \left(L^2 + \frac{L^4}{L_c^2}\right)\theta(L) + L^2\left[\phi^2 + \frac{L^2}{L_m^2}(\phi^2 + a\phi^4)\right]\theta(L), \quad (4.148)$$

where $L_m = 1/\omega$, $\lambda/4! = a/L_m^2$ and the PI measure $\mu = \exp(-L^4/L_0^4)$.

The first-order quantum correction to the classical action (4.141) is determined by

$$\Gamma_1 = i\frac{V_4}{L_0^4} + \frac{i}{2}Tr\log\begin{pmatrix} S_{LL} & S_{L\phi} \\ S_{L\phi} & S_{\phi\phi} \end{pmatrix}, \quad (4.149)$$

where $S_{xy}$ are the submatrices of the Hessian matrix for $S_{Rm}$. Since

$$S_{LL} = O(L^2), \quad S_{L\phi} = O(L^3)O(\phi), \quad S_{\phi\phi} = O(L^4)[1 + O(\phi^2)], \quad (4.150)$$

for $L$ large, then

$$\Gamma_1 = i\frac{V_4(L)}{L_0^4} + \frac{i}{2}Tr\log S_{LL} + \frac{i}{2}Tr\log S_{\phi\phi} + O(\phi^2). \quad (4.151)$$

The first term in (4.151) is the QG correction to the classical CC, while the matter sector will give a quantum correction to CC from the third term. This can be seen by considering the smooth manifold approximation, i.e. when $E \gg 1$. In this case the third term in (4.151) can be calculated by using the continuum approximation

$$S_{Rs}(L,\phi) \approx S_s(g_{\mu\nu},\phi), \quad (4.152)$$

and the corresponding QFT in curved spacetime, see section 4.6.

Let us consider an edge-length configuration which satisfies (4.132). The condition (4.132) ensures that the QG corrections are small and if $L_K \ll L_m$, we can calculate $Tr\log S_{\phi\phi}$ by using the Feynman diagrams for $S_s$ with the UV momentum cutoff $\hbar/L_K = \hbar K$. Consequently the corresponding CC contribution will be given by the flat space vacuum energy density, since

$$Tr\log S_{\phi\phi}\big|_{\phi=0} \approx V_M\int_0^K k^3\,dk\,\log(k^2 + \omega^2) + \Omega_m(R,K) \equiv \delta\Gamma_1(L), \quad (4.153)$$

and

$$\begin{aligned} \Omega_m(R,K) = \;& a_1 K^2 \int_M d^4x\sqrt{|g|}\,R \\ & + \log(K/\omega)\int_M d^4x\sqrt{|g|}\,[a_2 R^2 + a_3 R^{\mu\nu}R_{\mu\nu} \\ & \qquad\qquad + a_4 R^{\mu\nu\rho\sigma}R_{\mu\nu\rho\sigma} + a_5\nabla^2 R] \\ & + O\left(L_K^2/L^2\right), \end{aligned} \quad (4.154)$$

where $a_k$ are constants, $\hbar K$ is the momentum cutoff and $\hbar\omega$ is the mass of the scalar field, see [22]. Therefore the only $O(L^4)$ term in $\delta\Gamma_1$ is

$$c_1 V_M K^4 \log(K/\omega) = c_1 \frac{V_M}{L_K^4} \log(L_m/L_K), \qquad (4.155)$$

where $c_1$ is a numerical constant.

The physical effective action is given by the formula (4.125), so that the one-loop CC is given by

$$\Lambda_1 = \pm \frac{1}{2L_c^2} + \Lambda_\mu + c_1 \frac{l_P^2}{2L_K^4} \log(K/\omega), \qquad (4.156)$$

where $c_1$ is a numerical constant of $O(1)$. We can write this as

$$\Lambda_1 = \Lambda_\mu + \Lambda_c + \Lambda_m, \qquad (4.157)$$

and it is not difficult to see that the higher-loop matter contributions to CC will preserve this structure, due to (4.145) and (4.146). In section 5.3 we will give a detailed demonstration of this. Consequently

$$\Lambda = \Lambda_\mu + \Lambda_c + \Lambda_m, \qquad (4.158)$$

where

$$\Lambda_m = f(\lambda, \omega, l_P^2). \qquad (4.159)$$

The function $f$ is determined by the EA equation, see section 5.3, and $\lambda$ is the coupling constant of the scalar field self-interaction coupling constant. Note that the dependence on the cutoff $K$ has disappeared from $\Lambda_m$, because the exact value of $\Lambda$ comes from the solution of the EA equation, and this solution does not have any cutoff.

We can then choose the free parameter $L_c$ such that

$$\Lambda_c + \Lambda_m = 0, \qquad (4.160)$$

so that

$$\Lambda = \Lambda_\mu = \frac{l_P^2}{2L_0^4}. \qquad (4.161)$$

Note that $\Lambda_\mu > 0$ if we choose the $+$ sign in (4.125).

By taking $L_0 \approx 10^{-5} \, m$ we obtain the observed value of CC, which is

$$l_P^2 \Lambda \approx 10^{-122}. \qquad (4.162)$$

We obtained the same value for $L_0$ as in the pure gravity case, but now the consistency condition for the validity of the semiclassical approximation is not $L_0 \gg l_P$, but it is the condition (4.127). It implies that $|\Lambda_m| \gg \Lambda/2$, see section 5.3. If the value of $\Lambda_m$ violates the SC condition, one can adopt a different scheme of determining the free parameters. For example, by using $\Lambda = \Lambda_c$ and $\Lambda_\mu + \Lambda_m = 0$, the SC condition implies $|\Lambda_m| \ll \Lambda/2$.

These considerations are important because the value of CC can be measured only in the semiclassical regime of a Regge QG theory.

## 4.6   Smooth manifold approximation

In order to understand the effects of the Regge effective action for a smooth spacetime, we need to see how to approximate the Regge effective action with a QFT effective action.

Let $T(M)$ have a large number of the edges ($N \gg 1$) and let the variation of the edge lengths from each 4-simplex to its neighbor be small, i.e.

$$\big| |L_\epsilon(\sigma)| - |L_{\epsilon'}(\sigma')| \big| \leqslant \frac{l}{N} \,,$$

where $\sigma \cap \sigma' = \tau$ and $l$ a constant.

Given a function $f(L)$, we would like to approximate it with a functional of a smooth metric on $M$. The smooth limit can be defined as the limit $N \to \infty$ and $L_\epsilon \to 0$ such that

$$g_{\mu\nu}^{(\sigma)}(L) \to g_{\mu\nu}(x) \,,$$

where $x$ are the coordinates of a point inside the 4-simplex $\sigma$ and the partial derivatives of $g_{\mu\nu}(x)$ are continuous on $M$ up to order $n \geqslant 2$.

In the case of the Regge action, for large $N$ and a small local variation of the edge lengths, there is a smooth metric on $M$ such that

$$S_R(L) \approx \frac{1}{2} \int_M d^4x \sqrt{|g|} \, R(g) \,. \tag{4.163}$$

We also have

$$\Lambda_c V_4(L) \approx \Lambda_c \int_M d^4x \sqrt{|g|} = \Lambda_c V_M \,, \tag{4.164}$$

where $|g| = |\det g|$. These are the standard formulas of the Regge calculus and they nicely illustrate how functions $f(L)$ on the PL manifold $T(M)$ can be approximated by functionals of a smooth (differentiable) metric $g$ on $M$ for $N \to \infty$.

Similarly, the effective action $\Gamma(L)$ can be approximated by an effective action $\tilde{\Gamma}[g(x)]$ for a QFT with a momentum cutoff, where $g(x)$ is a smooth metric on $M$. Let

$$L_\epsilon = L_{0\epsilon} + l_\epsilon \,, \tag{4.165}$$

where $L_{0\epsilon}$ corresponds to a flat metric on $M$, such that

$$|L_{0\epsilon}| \approx L_0 \,, \quad |l_\epsilon| \ll L_0 \,, \quad L_0 \gg l_P \,.$$

Then

$$\Gamma(L) \approx \tilde{\Gamma}_{K_0}[\eta_{\mu\nu}(x) + h_{\mu\nu}(x)] \,, \tag{4.166}$$

where $g_{\mu\nu} = \eta_{\mu\nu} + h_{\mu\nu}$ and the QFT cutoff $K_0$ is proportional to $1/L_0$.

We saw previously that we need $L_\epsilon \gg l_P$ in order to have a perturbative semiclassical expansion, which may seem incompatible with the requirement $L_\epsilon \to 0$. However, this will not be a problem, since $l_P$ is extremely small, so that a microscopically small $L_\epsilon$ can be still much larger than $l_P$.

An example for the approximation (4.166) is

$$Tr \log S_R''(L) \approx \int_M d^4x \sqrt{|g|} \left(aR^2 + bR_{\mu\nu}R^{\mu\nu} + \cdots\right) \ln \frac{K_0}{k_0} \qquad (4.167)$$

where $g_{\mu\nu} = \eta_{\mu\nu} + h_{\mu\nu}$, $a$ and $b$ are numerical constants, while $k_0$ is an arbitrary constant such that $k_0 \ll K_0$. The $\cdots$ indicate some additional terms which may be present in the QFT effective action, like

$$c\,R \log \left(\frac{\Box}{m_0^2}\right) R + d\,R_{\mu\nu} \log \left(\frac{\Box}{m_0^2}\right) R^{\mu\nu},$$

see [17], where $\Box = g^{\mu\nu}\nabla_\mu\nabla_\nu$ and $m_0 = 1/k_0$.

The approximation (4.166) follows from the fact that a PL function on a lattice with a cell size $L_0$ can be written as a Fourier integral over a compact region $|q| \leqslant 2\pi/L_0$ where $q$ is the wave vector, see Appendix G. Hence the PL trace-log term can be approximated by the one-loop QFT effective action for GR by using a momentum cutoff $\hbar K_0 = 2\pi\hbar/L_0$.

Beside the standard trace-log term, the first-order effective action will also contain the $\log \mu$ term, so that

$$\Gamma_1(L) = Tr(\log S_R''(L)) + \frac{V_4(L)}{L_0^4} - p\sum_\epsilon \ln\left(1 + \frac{L_\epsilon^2}{l_0^2}\right), \qquad (4.168)$$

where we have discarded the constant term $-pN \ln l_0^2$.

The last term in (4.168) cannot be expressed as a functional of a metric in the smooth-manifold approximation. However, one can argue that it becomes negligibly small when $N \to \infty$ and $L_\epsilon \to 0$. Let $L_\epsilon \approx l_0/N$, then

$$\sum_\epsilon \ln\left(1 + \frac{L_\epsilon^2}{l_0^2}\right) \approx N \ln\left(1 + \frac{1}{N^2}\right) \approx \frac{1}{N}.$$

Hence the edge-length factors in the measure will be negligible in the smooth-manifold approximation. However, when $L_\epsilon \to \infty$, the edge-length terms will be subdominant in $\Gamma_1$ with respect to the $V_4(L)$ term, since $V_4(L) = O(L^4)$ while the edge-length terms behave as $O(\ln L)$. Therefore the measure $\mu_p$ will have the same semiclassical properties as the measure $\mu_0$.

# Chapter 5

# Applications of PLQG

## 5.1 Cosmological constant

We have explained in section 4.5 how the effective CC arises in the Regge QG model and in this section we will explain in more detail how the effective CC can be related with the observed value.

The effective CC in the Regge QG model is given as

$$\Lambda = \Lambda_c + \Lambda_{qg} + \Lambda_m \,, \tag{5.1}$$

where $\Lambda_c$ is a free parameter of the classical theory, $\Lambda_{qg}$ is given by (4.94) and $\Lambda_m$ is the matter contribution. From the QFT approximation we have

$$\Lambda_m \approx \sum_\gamma v(\gamma, K) \tag{5.2}$$

where $v(\gamma, K)$ is a one-particle irreducible vacuum Feynman diagram for the field-theory action $S_m$ in flat spacetime with the cutoff $K$ ($\hbar K$ is the 4-momentum cutoff).

In the case of a massive scalar field with a quartic self-interaction $\lambda$ one can show that

$$\sum_\gamma v(\gamma, K) \approx l_P^2 K^4 \left[ c_1 \ln(K^2/\omega^2) + \sum_{n \geqslant 2} c_n (\bar{\lambda})^{n-1} (\ln(K^2/\omega^2))^{n-2} \right.$$
$$\left. + \sum_{n \geqslant 4} d_n (\bar{\lambda})^{n-1} (K^2/\omega^2)^{n-3} \right] \,, \tag{5.3}$$

for $K \gg \omega$, where $\bar{\lambda} = l_P^2 \lambda$, $m = \hbar \omega$, see Appendix E. The infinite series in (5.3) will be divergent even for finite values of $K$, so that one has to apply some type of renormalization procedure for the effective QFT valid below the cutoff $K$.

If one uses dimensional regularization,[1] then at the one-loop order one obtains

$$\Lambda'_m \approx \hbar \sum_k (-1)^{f_k} m_k^4 \ln\left(\frac{m_k^2}{\mu^2}\right) \tag{5.4}$$

see [79], where $m_k$ are the masses of the SM particles and $f_k = 0$ for a boson and $f_k = 1$ for a fermion. The parameter $\mu$ is a relevant mass scale, so that $\mu^2 \leqslant K^2$.

One also has to add to (5.4) the contributions of the massless bosons, i.e. the photon and the gluons, although it can be argued that their contribution is zero, because $m_k = 0$, see [79]. However, as far as the graviton is concerned, the corresponding QFT is non-renormalizable, so that the vacuum energy density cannot be removed. Consequently we have to add to $\Lambda'_m$

$$\Lambda'_{qg} \approx \pm\hbar K^4 = \pm\frac{l_P^2}{L_K^4}. \tag{5.5}$$

In the QFT approach one also has the relationship

$$\Lambda = \Lambda_c + \Lambda'_{qg} + \Lambda'_m. \tag{5.6}$$

If we take the natural cutoff $L_K = l_P$, then (5.6) gives

$$\Lambda \approx \Lambda_c \pm \frac{1}{l_P^2}, \tag{5.7}$$

since $\Lambda'_{qg} \gg |\Lambda'_m|$. We can rewrite the equation (5.7) as

$$l_P^2 \Lambda \approx l_P^2 \Lambda_c \pm 1, \tag{5.8}$$

and given that the observed value of the cosmological constant satisfies $l_P^2 \Lambda \approx 10^{-122}$, then we obtain

$$10^{-122} \approx \lambda_c \pm 1,$$

where $\lambda_c = l_P^2 \Lambda_c$. Therefore $|\lambda_c|$ has to be different from 1 at the 122-nd decimal place.

This extreme fine tuning of the parameter $\Lambda_c$ is the famous aspect of the cosmological constant problem[2] which appears in the QFT formulation of quantum gravity. If the relation (5.7) were an exact relationship, than any value for $\Lambda_c$ would be acceptable. However, the relation (5.7) is only a comparison of the orders of magnitude and the corresponding extreme fine

---

[1] Dimensional regularization can be related to the cutoff regularization.
[2] See the next section for the exact definition of the CC problem.

tuning tells us that it cannot be accepted even as an approximate formula for the CC.

However, in the PL formulation of quantum gravity, the QFT which produces the extreme fine tuning is just an approximation. The fundamental theory has finitely many DoF, so that the exact solution of the EA equation will give a finite and cutoff-independent value for $\Lambda$. Therefore

$$\Lambda_m = f(\vec{\omega}, \vec{\lambda}, l_P^2), \tag{5.9}$$

where $\vec{\omega} = (\omega_1, \ldots, \omega_n)$ and $\vec{\lambda} = (\lambda_1, \ldots, \lambda_r)$ are the masses ($m = \hbar\omega$) and the couplings of the Standard Model and $f$ is the function determined by the EA equation (4.143). Consequently we obtain an exact (i.e. non-perturbative) relation

$$\Lambda = \pm\frac{1}{L_c^2} + \frac{l_P^2}{2L_0^4} + f(\vec{\omega}, \vec{\lambda}, l_P^2). \tag{5.10}$$

Equation (5.10) can be used to fix the free parameters $L_0$ and $L_c$. By equating $\Lambda$ with the experimentally observed value, we obtain

$$\lambda_o = x + y + \lambda_m \tag{5.11}$$

where $\lambda_o = l_P^2\Lambda \approx 10^{-122}$, $x = \pm l_P^2/L_c^2$, $y = l_P^4/2L_0^4$ and $\lambda_m = l_P^2 f$. The equation (5.11) has infinitely many solutions, but we also have to impose the condition for the existence of the semiclassical limit (4.127). This gives the restriction

$$0 < y \ll 2|x|. \tag{5.12}$$

The value of $\lambda_m$ is not known, but for any value of $\lambda_m$ the equation (5.11) has infinitely many solutions which obey the restriction (5.12). Note that the solution $x = -\lambda_m$ and $y = \lambda_o$, which was proposed in [99], will be acceptable if $|\lambda_m| \gg \lambda_o/2$. This solution is special because it gives a value for $L_0$ which is independent of the value of $\lambda_m$, specifically $L_0 \approx 10^{-5} m$. This is the same value which was obtained in the case of pure PL gravity without the cosmological constant [88]. Another possibility is a choice $\Lambda_c = 0$, which gives $\Lambda = \Lambda_\mu + \Lambda_m$ and the SC criterion is $L_0 \gg l_P$. This implies

$$\Lambda \ll \Lambda_m + \frac{1}{2l_P^2}.$$

Therefore, whatever is the value of $\lambda_m$, there is a choice of the parameters which are consistent with the restrictions imposed by the validity of the SC approximation.

### 5.1.1 *The CC problem in quantum gravity*

The formula (5.10) for the exact effective cosmological constant is an essential ingredient for the resolution of the CC problem from QFT in the context of a QG theory. The result (5.10) can be better understood if we recall the definition of the CC problem given by Polchinski [107]. According to this definition, the CC problem in a QG theory has two parts:

(P1) show that the observed CC value lies in the spectrum of the CC,
(P2) explain why the CC takes the observed value.

The meaning of the first part (P1) of the CC problem is obvious if the cosmological constant is represented by an operator. In the case when one has a quantum corrected expression of the classical CC value, one has to show that there are values of the free parameters which give the observed CC value. The PLQG theory clearly solves (P1), while the second part (P2) of the CC problem cannot be addressed by the current formalism of quantum theory. The reason is that one has to generalize the standard formalism of quantum mechanics in order to provide a mechanism for a selection of a wavefunction of the universe with a particular value of the cosmological constant.

Note that demonstrating (P1) is a highly non-trivial task in any QG theory. The problem (P1) has been addressed so far only in PLQG theory and in string theory. In the string theory case there are only plausibility arguments that (P1) is true [27, 68]. The CC spectrum in string theory is discrete with $O(10^{500})$ values [27]. Although positive CC values are not natural in string theory, a mechanism for their appearance was provided in [68]. Hence it is plausible to assume that the CC spectrum is sufficiently dense around zero such that the observed value is sufficiently close to some CC spectrum value.

The part (P2) of the CC problem has been only addressed in string theory. This is the multiverse proposal, see [120], and the assumption is that there are many universes, each having some fixed CC value from the CC spectrum. According to this proposal, we happen to live in the universe with the CC value $\Lambda_c l_P^2 \approx 10^{-122}$, because this is the value that allows formation of galaxies, planets and life, see [127] for the anthropic determination of the CC value.

Note that there are many proposals for (P2) which are not derived from a QG theory, but instead they assume that a certain effective action exists such that its EoMs give the required CC value, see for example [40].

## 5.2   Quantum cosmology

In canonical quantization of GR spacetime $M$ must have the "slab" topology $\Sigma \times I$, where the interval $I$ is a subset of $\mathbb{R}$, and the spacetime metric is given by

$$ds^2 = -(N^2 - n^i n_i)dt^2 + 2n_i\, dt dx^i + h_{ij}\, dx^i dx^j\,, \qquad (5.13)$$

where $(x^i, t)$ are coordinates on $\Sigma \times I$, $N$ is the lapse and $n^i$ is the shift vector, while $h_{ij}$ is a metric on $\Sigma$. The canonical analysis of the AH action for the metric (5.13) gives that the canonical variables are $h_{ij}$ and its canonically conjugate momenta $p_h{}^{ij}$, which are constrained by the diffeomorphism constraints $D_i(p_h, h)$ and the Hamiltonian constraint $W(p_h, h)$, see section 1.2. The canonical quantization then gives that a wavefunction $\Psi(h)$ has to be invariant under the 3-diffeomorphisms of $\Sigma$ and $\Psi(h)$ has to obey the Wheeler-DeWitt (WDW) equation

$$\hat{W}(\hat{p}_h, \hat{h})\Psi(h) = 0\,, \qquad (5.14)$$

where $\hat{W}$ is an operator obtained by substituting the variables in the function $W(p_h, h)$ with the operators $\hat{p}_h{}^{ij}$ and $\hat{h}_{ij}$, see section 1.2.

Solving the WDW equation (5.14) in general case is notoriously difficult. However, Hartle and Hawking have proposed a way to construct a solution which describes an initial wavefunction of the universe [60]. It is given by a path integral

$$\Psi_0(h) = \int \mathcal{D}g\, e^{-\int_M d^4 x \sqrt{g}(R(g)+\lambda)/l_P^2}\,, \qquad (5.15)$$

where a spacetime $M$ has the topology of a cup such that $\partial M = \Sigma$ and the metrics $g$ have the Euclidean signature such that $g|_{\partial M} = h$.

Even the Hartle-Hawking (HH) wavefunction $\Psi_0(h)$ can be calculated only in some special cases. These are the minisuperspace models where the spacetime metric has a finite number of DoF. For example, the Friedmann-Lemaître-Robertson-Walker (FLRW) metric is given by

$$ds^2 = -N^2(t)\, dt^2 + a^2(t)\, (dx^2 + dy^2 + dz^2)\,, \qquad (5.16)$$

and there are two DoF, the scaling factor $a$ and the lapse $N$. Consequently

$$\Psi_0(a) = \int_J dN \int \mathcal{D}a \exp\left(-\int_I dt\, L_E(a, \dot{a}, N)/l_P^2\right)\,, \qquad (5.17)$$

in the gauge $N(t) = \text{const}$, where $J, I \subseteq \mathbb{R}$ and $L_E$ is the Euclidean metric extension of the Lagrangian $L(a, \dot{a}, N)$ corresponding to the metric (5.16).

For a general minisuperspace model the path integral (5.17) can be calculated only approximately by using the stationary phase approximation. Also, in order to obtain a solution of the WDW equation the interval $J$ has to be promoted into a contour in the complex plane, see [57, 58]. Even in the Lorentzian version of (5.17), one has to extend $J$ into a complex plane contour [46].

### 5.2.1   Hartle-Hawking wavefunction

The PLQG formulation also offers a possibility to calculate the HH wave-function, since one can mimic the minisuperspace models by using simple triangulations where many of the edge lengths are the same. Also there is an advantage that the spacetime geometry is transparent in the PL case, so that all the domains of integration are uniquely determined and there is no need for complex domains of integration since the convergence is achieved through the PI measure (3.109).

For example, let us consider the case when $M = S^4$ (a 4-sphere) such that $\partial M = \Sigma = S^3$ (a three-sphere). We will then consider a triangulation $T(S^4) = \sigma_6$ (a 4-dimensional simplicial complex based on six points), which we embed in $\mathbb{R}^5$, with $T(S^3) = \sigma_5$ (a 3-dimensional subcomplex of $\sigma_6$ based on 5 points), such that $L_\epsilon = l > 0$ for $\epsilon \in \sigma_5$, $L_\epsilon = s > 0$ for $\epsilon \in \sigma_6 \setminus \sigma_5$. Consequently

$$S_R(l, s) = \frac{5\sqrt{3}}{2} l^2 \delta_1(l, s) + \frac{5}{2} l \sqrt{s^2 - \frac{l^2}{4}} \, \delta_2(l, s), \qquad (5.18)$$

where

$$\delta_1 = 2\pi - 2\alpha, \quad \delta_2 = 2\pi - 3\beta, \qquad (5.19)$$

and

$$\sin \alpha = \frac{\sqrt{s^2 - \frac{3l^2}{8}}}{\sqrt{s^2 - \frac{l^2}{3}}}, \quad \sin \beta = \frac{2\sqrt{2}\sqrt{s^2 - \frac{3l^2}{8}}\sqrt{s^2 - \frac{l^2}{4}}}{3(s^2 - \frac{l^2}{3})}. \qquad (5.20)$$

The HH path integral is then given by

$$\Psi_0(l) = \int_{l_1}^{\infty} ds \, \mu(l, s) \exp\left(-S_R(l, s)/l_P^2\right), \qquad (5.21)$$

where $l_1 = \sqrt{\frac{3}{8}} \, l$ and

$$\mu(l, s) = \exp\left(-\frac{5\sqrt{2}\, l^3}{48 \, L_0^4} \sqrt{s^2 - \frac{3l^2}{8}}\right) \left(1 + \frac{s^2}{l_0^2}\right)^{5p/2}. \qquad (5.22)$$

We do not include the edge-length measure factor for the edges of length $l$, since they are not integrated over.

The integral (5.21) is convergent because when $s \to +\infty$ we have $\alpha \to \pi/2$, $\beta \to \arcsin(2\sqrt{2}/3)$ so that

$$\mu \, e^{-S_R/l_P^2} \approx e^{-\frac{5\sqrt{3}\pi}{4} l^2/l_P^2} \exp\left(-(\lambda_0 \, l^2 + \delta) l s/l_P^2\right), \qquad (5.23)$$

where $\lambda_0 = 5\sqrt{2}\, l_P^2/(48 L_0^4)$ and $\delta = 2\pi - 3\arcsin(2\sqrt{2}/3)$. Since $s$ is large, we can neglect the contribution $5p \ln s$ in the exponent of (5.23), which is a linear function of $s$.

The HH integral is also convergent for the trivial measure $\mu = 1$, which corresponds to $\lambda_0 = 0$. However, when a cosmological constant term is included, the HH integral will be divergent for a sufficiently negative $\lambda$, since then

$$\mu \, e^{-S_R/l_P^2} \approx e^{-\frac{5\sqrt{3}\pi}{4} l^2/l_P^2} \exp\left(-[(\lambda_0 + \lambda)l^2 + \delta] l s/l_P^2\right), \qquad (5.24)$$

and $(\lambda_0 + \lambda)l^2 + \delta < 0$. This illustrates the earlier observation that a Euclidean GR path integral is not always convergent.

A natural question arises in relation to the PL HH wavefunction (5.21), and that is whether it satisfies a WDW equation. Note that the scale factor and the lapse are given by

$$a = l/l_0, \quad N = s/t_0, \qquad (5.25)$$

where $l_0$ is a unit of length and $t_0$ is a unit of time. One can then ask is there a WDW operator $\hat{W}$ such that $\hat{W}\Psi_0(a) = 0$? More precisely, are there some constants $\alpha, \beta$ and $\gamma$ such that

$$\hat{W}\Psi_0 = \alpha \frac{1}{a} \frac{d^2\Psi}{da^2} + \beta \frac{d}{da}\left(\frac{1}{a}\frac{d\Psi}{da}\right) + \gamma \frac{d^2}{da^2}\left(\frac{\Psi_0}{a}\right) + \lambda a^3 \Psi_0 = 0. \qquad (5.26)$$

The equation (5.26) does not necessarily hold in PLQG, because the WDW equation corresponds to a smooth manifold $M$, while we have a PL manifold $T(M)$. However, when $N_1 \to \infty$, i.e. in the smooth-manifold limit, we expect that

$$\hat{W}_{T(M)} \to \hat{W}_M, \qquad (5.27)$$

where $\hat{W}_{T(M)}$ is an operator such that $\hat{W}_{T(M)}\Psi_0(a) = 0$.

The bosonic matter can be coupled to gravity by using the Euclidean PL metric (2.3). In the case of a scalar field $\phi$ we have

$$S_m = \sum_{\sigma \in T(M)} V_\sigma \mathcal{L}_\sigma + \sum_{\pi \in T(M)} V_\pi^*(L) U(\pi), \qquad (5.28)$$

where

$$\mathcal{L}_\sigma = \frac{1}{2} g^{\mu\nu}(\sigma) \, \Delta_\mu \phi \, \Delta_\nu \phi \tag{5.29}$$

and $U(\phi)$ is the scalar field potential. The metric $g^{\mu\nu}(\sigma)$ is given by the inverse matrix of (2.3), while

$$\Delta_\mu \phi = \frac{\phi_\mu - \phi_0}{L_{0\mu}}, \tag{5.30}$$

where $\phi_\mu = \phi(\xi_\mu)$ and $\phi_0 = \phi(\xi_0)$, while $\xi_0$ and $\xi_\mu$ are the vertices of $\sigma$, see section 2.2.

In the case of the triangulation corresponding to the Regge action (5.18), we have the following HH wavefunction

$$\Psi(l, f) = \int_{l_1}^{\infty} ds \int_{-\infty}^{\infty} d\varphi \, \mu(l, s) \exp\left(-\frac{1}{l_P^2} \left[S_R(l, s) - S_m(l, s, \varphi, f)\right]\right), \tag{5.31}$$

where $\phi(\xi) = f$ for $\xi \in \sigma_5$ and $\phi(\xi) = \varphi$ for $\xi \in \sigma_6 \setminus \sigma_5$. Note that in the Euclidean case the matter action is $-S_m$. A necessary but not sufficient condition for the convergence of (5.31) is that the integral

$$\int_{-\infty}^{\infty} d\varphi \exp\left(-\frac{1}{l_P^2} S_m(l, s, \varphi, f)\right), \tag{5.32}$$

is convergent. For the polynomial potentials $U(\phi)$ this requires that $U(\phi) > 0$ for $\phi \to \pm\infty$.

The fermionic matter can be coupled by using the PL tetrads $e_\mu^a(\sigma)$, defined by

$$e_\mu^a(\sigma) \, e_\nu^b(\sigma) \, \eta_{ab} = g_{\mu\nu}(\sigma), \tag{5.33}$$

and the PL spin connection $\omega_\mu^{ab}(\epsilon^*)$, where $\epsilon^*$ is the dual edge connecting the centers of two adjacent 4-simplices, see section 2.2.

### 5.2.2  *Vilenkin wavefunction*

We saw in the previous section that the HH path integral for the PL version of the FLRW minisuperspace model was not always convergent. This is a manifestation of a more general problem that a Euclidean path integral is not always convergent, because, beside the sign of the cosmological constant, the sign of the scalar curvature and the interval of its values affect the convergence of the path integral. One can remedy this situation by using the Minkowski signature metrics, but as we saw in chapter 4, the definition of a Minkowski path integral requires the spacetime topology of a slab or a cylinder, which is different from the cup topology which was used to define the HH path integral. However, one can simulate the cup topology by collapsing the cylinder to a cone, which is essentially the Vilenkin proposal for the wavefunction of the universe, see [122].

Note that the path integral (3.101) is a function of the initial edge lengths $L_\epsilon = l'_\epsilon$ on $T_0(\Sigma)$ and the final edge lengths $L_\epsilon = l_\epsilon$ on $T_n(\Sigma)$. This function is known as the propagator, which we denote as $G(l, l')$. We denote as $G(h, h')$ the smooth-manifold version of $G(l, l')$, where $h$ is the metric on the final boundary $\Sigma$ and $h'$ is the metric on the initial boundary $\Sigma$. The smooth propagator satisfies a non-homogeneous WDW equation

$$\hat{W}(\hat{p}_h, \hat{h})\, G(h(x), h'(y)) = \prod_{y \in \Sigma} \delta(h(x) - h'(y))\,. \tag{5.34}$$

The Vilenkin proposal for the wavefunction of the universe is to take $h' = 0$, so that $G(h, 0)$ satisfies the WDW equation for $h \neq 0$, see [122]. In the PL case, the analog of the Vilenkin wavefunction will be the propagator $G(l, 0)$, and it is easy to see that it will be identical to the Lorentzian version of the HH wavefunction for the conical triangulation.

Let us consider 6 points in $\mathbb{R}^5$ such that

$$v_{jk}^2 = l^2\,, \quad v_{j5}^2 = 3l^2/8 - t^2 \equiv l_1^2 - t^2\,, \quad t \in [0, \infty)\,, \tag{5.35}$$

where $j, k = 1, 2, \ldots, 4$ and $j \neq k$. This corresponds to an embedding of $\sigma_6$ into $\mathbb{R}^5$ with coordinates $(x, y, z, w, t)$ such that $\sigma_5$ is embedded into the spatial hyperplane $t = 0$ with the point $(0, 0, 0, 0, 0)$ corresponding to the center of the 3-sphere which contains the spatial points $(x_j, y_j, z_j, w_j, 0)$ such that $d_{ij} = l$. Note that

$$v_{j5} = \begin{cases} s & t \leqslant l_1 \\ is & t > l_1, \end{cases} \tag{5.36}$$

where

$$s = \sqrt{|l_1^2 - t^2|} \tag{5.37}$$

and $l_1 = \sqrt{3l/8}$ is the radius of the 3-sphere. We will then define the Lorentzian HH integral as

$$\Psi_0(l) = \int_0^{l_1} ds\, \mu(s, l)\, e^{i(\tilde{S}_R(s, l) + \lambda V_4)/l_P^2} + \int_0^\infty ds\, \mu(s, l)\, e^{i(\tilde{S}_R(s, l) + \lambda V_4)/l_P^2}\,, \tag{5.38}$$

where

$$S_R(l, s) = \frac{5\sqrt{3}}{2} l^2\, \delta_1(l, s) + \frac{5}{2} l \sqrt{\left| s^2 - \frac{l^2}{4} \right|}\, \delta_2(l, s) \tag{5.39}$$

and $\tilde{S}_R$ is given by (2.26), while

$$\delta_1 = 2\pi - 2\alpha, \quad \delta_2 = 2\pi - 3\beta, \tag{5.40}$$

and

$$V_4(l, s) = \frac{5\sqrt{2}}{48} l^3 \sqrt{\left| s^2 - \frac{3}{8} l^2 \right|}. \tag{5.41}$$

The expressions for $\alpha$ and $\beta$ will be given by the definition (2.22) and by the embedding (5.35). In order to see what is $\tilde{S}_R$, it is helpful to write the dihedral angles in terms of the $t$ variable

$$\sin \alpha = \begin{cases} \frac{it}{\sqrt{l_3^2 - t^2}} = i \sinh a, \, t < l_3 \\ \frac{t}{\sqrt{t^2 - l_3^2}} = \cosh a, \, \, t > l_3 \end{cases} \tag{5.42}$$

and

$$\sin \beta = \begin{cases} \frac{2\sqrt{2}}{3} \frac{it\sqrt{l_2^2 - t^2}}{l_3^2 - t^2} = i \sinh b, \, t < l_3 \\ \frac{2\sqrt{2}}{3} \frac{t\sqrt{t^2 - l_2^2}}{t^2 - l_3^2} = \sin b, \, \, \, \, t > l_3 \end{cases}, \tag{5.43}$$

where $l_2 = l/\sqrt{8}$ and $l_3 = l/\sqrt{24}$. Note that these expressions can be obtained from the Euclidean expressions (5.20) by performing an analytic continuation $t \to it$ (a Wick rotation).

The convergence of the Lorentzian HH integrals (5.38) reduces to the convergence of the second integral. This integral is convergent due to the large-$s$ asymptotics

$$\mu\, e^{iS_R} \approx C\, e^{-\lambda_0 l^3 s/l_P^2 + 5p \ln s}\, e^{i(\lambda l^2 + \delta)ls/l_P^2} \approx C\, e^{-\lambda_0 l^3 s/l_P^2}\, e^{i(\lambda l^2 + \delta)ls/l_P^2}. \tag{5.44}$$

As in the Euclidean case, the contribution $5p \ln s$, which comes from the edge-length factor in the measure, can be neglected for large $s$. This means that in the case of minisuperspace models, one can take the $p = 0$ path-integral measure.

### 5.2.3   *Additional remarks*

The Vilenkin PL wavefunction (5.38) is defined for any value of the cosmological constant $\lambda$, while the Hartle-Hawking PL wavefunction (5.17) is defined only for

$$\lambda > -\lambda_0 - \frac{\delta}{l^2}. \tag{5.45}$$

This reflects the fact that the Lorentzian path integral (3.101) has better convergence properties than the Euclidean path integral (3.97). Also, the

Vilenkin PL wavefunction is the same as the Lorentzian version of the HH PL wavefunction, a fact which indicates that this may be true in the smooth limit approximation.

Note that the propagator $G(h, h')$ is not the same as the Schrödinger equation propagator $K(\tilde{h}, T; \tilde{h}', T')$, where $K$ satisfies the deparametrized form of the WDW equation

$$\left[ i\hbar \frac{\partial}{\partial T} - \hat{H}(p_{\tilde{h}}, \tilde{h}, T) \right] K(\tilde{h}, T; \tilde{h}', T') = 0 \,. \tag{5.46}$$

Namely, one can perform a deparametrization of the Hamiltonian constraint $W(p_h, h)$, which amounts to performing a canonical transformation $(p_h, h) \to (p_{\tilde{h}}, P_T; \tilde{h}, T)$ such that

$$W(p_h, h) = 0 \Leftrightarrow P_T + H(p_{\tilde{h}}, \tilde{h}, T) = 0 \,, \tag{5.47}$$

see section 1.2. Consequently, the canonical quantization with respect to the new canonical variables will give the Schrödinger equation (5.46) in the gauge $\partial_i T = 0$, i.e. a gauge choice[3] where $T$ does not depend on the spatial coordinates $x^i$.

For example, in the case of the relativistic particle, when the WDW equation becomes the KG equation, we have for the propagator

$$G(\vec{x}, t) = \int_{\mathbb{R}^4} d\omega \, d^3 \vec{k} \, \frac{e^{-i\omega t + i\vec{k} \cdot \vec{x}}}{\omega^2 - \vec{k}^2 - \omega_0^2} \,, \tag{5.48}$$

where $\omega = p_0/\hbar$, $\vec{k} = \vec{p}/\hbar$ and $\omega_0 = m/\hbar$, while for the Schrödinger propagator one obtains

$$K(\vec{x}, t) = \int_{\mathbb{R}^3} d^3 \vec{k} \, e^{-it\sqrt{\vec{k}^2 + \omega_0^2} + i\vec{k} \cdot \vec{x}} \,. \tag{5.49}$$

In the PL context, the Schrödinger propagator becomes $K(\tilde{l}_\epsilon, T_k; \tilde{l}'_\epsilon, T_{k'})$, where $T_k = k l_0$ is an PL analogue of the time variable $T$ and $k, k' \in \mathbb{N}$. The choice of the time variable can be implemented through the following restrictions

$$V(\Sigma_k) = f(k) \, l_0^3 \,, \quad k = 0, 1, 2, \ldots, n \,, \tag{5.50}$$

where $f(k)$ is a given function. The restrictions (5.50) have to be imposed in the path integral (3.101) in order to obtain the Schrödinger propagator $K$.

---

[3]The question of the existence of such a gauge choice is called the problem of time in canonical GR, see [65].

Note that any solution of a deparametrized WDW equation can be written as

$$\Phi(\tilde{h}, T) = \int \mathcal{D}\tilde{h}' \, K(\tilde{h}, T; \tilde{h}', 0) \, \Phi_0(\tilde{h}') \,, \qquad (5.51)$$

where $\Phi_0(\tilde{h}')$ is the initial wavefunction of the universe (WFU). The functional $\Phi_0(\tilde{h})$ is arbitrary, in contrast to the HH or the Vilenkin wavefunctions, which should satisfy the WDW equation $\hat{W}\Psi(h) = 0$. It looks like that there is a greater freedom in choosing a WFU in the Schrödinger framework than in the WDW framework, so that this question deserves a further study.

The study of the HH path integral with matter (5.31) and its Lorentzian version are obvious further steps in the PL approach to the problem of the WFU. Another interesting problem for further study would be a determination of the smooth limits of the PL Hartle-Hawking and the Vilenkin wavefunctions which can be done by using a conical spacetime triangulation where the spatial simplicial complex $\sigma_5$ is replaced by $\sigma_n$ and then let $n \to \infty$.

## 5.3 Other applications

### 5.3.1 *Time evolution of the universe*

Given an effective action $\Gamma(L, \phi)$, one would like to see what could be the quantum trajectories, i.e. the values for $L_\epsilon$ and $\phi_v$ which satisfy the quantum equations of motion

$$\frac{\partial \Gamma(L, \phi)}{\partial L_\epsilon} = 0, \quad \epsilon = 1, 2, \ldots, E \,,$$

and

$$\frac{\partial \Gamma(L, \phi)}{\partial \phi_v} = 0, \quad v = 1, 2, \ldots, V \,,$$

where $E$ is the number of the edges and $V$ is the number of the vertices in a triangulation $T(M)$.

As we have explained earlier, the effective action only makes sense for $M = \Sigma \times [t_i, t_f]$, where $t_i$ is the initial time, while $t_f$ is the final time. The time evolution in a solution set $\{L_\epsilon \,|\, \epsilon = 1, 2, \ldots, E\}$ can be then understood by using a casual triangulation defined in section 3.6, such that the time-like edges in a slice

$$T(M_k) = T\left(\Sigma \times [t_k, t_{k+1}]\right) \,,$$

have imaginary values, while the space-like edges have real positive values, where $k = 0, 1, 2, \ldots, n - 1$ and $t_0 = t_i$ while $t_n = t_f$. The space-like edges in $T(M_k)$ correspond to the edges of $T_k(\Sigma)$ and of $T_{k+1}(\Sigma)$.

We can then say that a set of space-like edge lengths $L_\epsilon^{(s)}(k)$ from $T_k(\Sigma)$ is a time evolution of the initial set of space-like edge lengths $L_\epsilon^{(s)}(0)$ from $T_0(\Sigma)$. Also one can associate a physical time interval between $T_k$ and $T_0$ as

$$\Delta t_{phys}(k) = \sum_{m=0}^{k} |L_t(m)| \,, \qquad (5.52)$$

where $L_t(k)$ is the average of time-like edge lengths in $T(M_k)$.

Let $\bar{L}_k$ denote the average of $|L_\epsilon|$ in $T(M_k)$. In the early universe we expect

$$\bar{L}_k \approx l_P \,,$$

while in the present universe we expect that

$$\bar{L}_k \gg l_P \,.$$

We also expect that for the present universe $\bar{L}_k$ are some microscopic lengths, which are smaller than the spacetime distance probed in the LHC experiments, which are of the order of $10^{-20}\,m$. Otherwise the PL structure would create significant deviations from the usual QFT predictions, and since no significant deviations have been seen in the experiments at this scale, this means that $\bar{L}_k$ are smaller than $10^{-20}\,m$. For example, the smooth-manifold approximation (4.166) is given by

$$\Gamma(L, \phi) \approx \tilde{\Gamma}_K[g_{\mu\nu}(x), \phi(x)] \,,$$

where $\tilde{\Gamma}$ is the effective action for the GR plus matter QFT with a cutoff $K$. We also have from (4.165) that

$$L_\epsilon = L_{0\epsilon} + l_\epsilon \approx \bar{L}_k + l_\epsilon \,,$$

in $T(M_k)$, so that

$$K \approx \frac{2\pi}{\bar{L}_k} \,. \qquad (5.53)$$

Since the QFT cutoff must be greater than the maximal collision energy in the LHC experiments, which is $13\,TeV$, we conclude that $\bar{L} < 10^{-20}\,m$. The restrictions on $\bar{L}$ imply

$$10^{16}\,TeV \gg \hbar K > 13\,TeV \,, \qquad (5.54)$$

where $10^{16}\,TeV$ is the Planck energy. This leaves the possibility that for the collision energies anywhere in the range from $100\,TeV$ to $10^{16}\,TeV$ one may detect the deviations from the QFT predictions which are due to the PL structure, since then one may probe the distances which are comparable to the average edge length.

### 5.3.2   Non-perturbative EA

The effective action equation can be solved perturbatively for $L_\epsilon \gg l_P$, which is the semiclassical regime. An important problem is how to solve the EA equation for $L_\epsilon \approx l_P$, which is the deep quantum regime. In this region the perturbation theory fails, and one has to find an alternative method. One expects that the early universe is in this regime.

A natural approach for the deep quantum regime is to use the fact that the effective action is also the generating functional for the one-particle-irreducible (1PI) Green's functions, so that

$$\Gamma(L) = \sum_{\epsilon,\epsilon'} \tilde{\Gamma}_2(\epsilon,\epsilon') L_\epsilon L_{\epsilon'} + \sum_{\epsilon,\epsilon',\epsilon''} \tilde{\Gamma}_3(\epsilon,\epsilon',\epsilon'') L_\epsilon L_{\epsilon'} L_{\epsilon''} + \cdots , \qquad (5.55)$$

where $\tilde{\Gamma}_n(\epsilon)$ is the 1PI part of the $n$-point Green's function

$$G(\epsilon_1,\ldots,\epsilon_n) = \frac{1}{Z} \int_D d^E L \,\mu(L)\, L_{\epsilon_1} \cdots L_{\epsilon_n}\, e^{iS_R(L)/l_P^2} , \qquad (5.56)$$

and $Z$ is the Regge path integral for $T(M)$. The Green's functions given by (5.56) can be calculated numerically, which can be then used to obtain the expansion (5.55).

One can also apply these techniques in the context of the PLQG mini-superspace approximation introduced in section 5.2. The calculation of the Green's functions could be then done analytically.

# Chapter 6

# PLQG and other QG models

## 6.1  Canonical quantization and PLQG

A connection between the canonical QG and PLQG can be made via the Hartle-Hawking or the Vilenkin wavefunctions, since these wavefunctions can be expressed as path integrals, while it is expected that both wavefunctions satisfy the WDW equation.

In the case of the HH wavefunctions, it was shown that a HH wavefunction satisfies the WDW equation only in the case of minisuperspace models, see section 5.2. When the path integral is over a more general class of metrics, there is a problem in calculating the HH path integral, since the convergence of a Euclidean Regge path integral is not guaranteed, see section 3.6.

In the case of a Vilenkin wavefunction, there is no problem with the convergence, since a Lorentzian Regge path integral over a general class of metrics can be made convergent, by choosing an appropriate path-integral measure, see section 3.6. Then one can expect that such a wavefunction will obey some PL form of the WDW equation, which should yield some normal-ordered form of the WDW equation in the smooth manifold limit.

As far as the EA is concerned, it is not immediately clear what is the relationship between these wavefunctions and the effective action. Note that the way we defined the effective action was independent from a choice of a wavefunction on the initial surface $\Sigma$. Since the path integral on the 4-cylinder $\Sigma \times [t_1, t_2]$ can be associated to the Vilenkin wavefunction when the metric on the initial surface tends to zero, then it is natural to associate the Vilenkin wavefunction to the Regge effective action defined by the equation (4.79).

In the case of a HH wavefunction, there is no unique HH wavefunction which can be associated to a 3-manifold $\Sigma$ with a metric $h$, so it is not clear which HH wavefunction could be associated to the EA. The reason is that there are infinitely many topologically inequivalent 4-manifolds whose boundary is $\Sigma$. Therefore when calculating a HH path integral, one fixes a triangulation which corresponds to some 4-manifold, and after the integration of the edge lengths, the result will depend on the topological class of $M$. Each topological class will generate a different HH wavefunction, and the only way to obtain a unique HH wavefunction is to sum over the topological classes, which is a hopeless task. In the PLQG context this would mean a sum over triangulations. Therefore, provided one solves the problem of the convergence of the HH path integrals, one could generate a lot of new solutions of the WDW equation.

Given that the Vilenkin wavefunction is unique for a given 3-manifold $\Sigma$, one can ask how to generate new physical wavefunctions. Since the Vilenkin wavefunction is a special case of the evolution operator kernel, see section 5.2.2, we can generate a new physical wavefunction by integrating the evolution kernel $K(\tilde{h}, T; \tilde{h}', 0)$ which acts on an arbitrary functional of $\tilde{h}'$, where $\tilde{h}$ are the physical DoF of the 3-metric $h$. The PL version of this path integral will give many new solutions of the PL WDW equation. However, it is not clear how to change the EA equation (4.79) such that it will correspond to a new initial state (5.51).

Note that Regge QG for a PL manifold $T(\Sigma \times [t_1, t_2])$ is more general than the canonical QG, since the PL version of the canonical QG corresponds to a special case when the triangulations $T_k(\Sigma)$, $k = 0, 1, 2, \ldots, n$, are all the same, see subsection 5.3.1. One can refer to this type of a triangulation as a Cauchy triangulation, since in the classical case there is a unique trajectory associated to an edge length in $T_0(\Sigma)$. Hence the semiclassical limit of a canonical QG will be described by the effective action for a Cauchy triangulation.

## 6.2   PLQG and Causal Dynamical Triangulations

In the Causal Dynamical Triangulations (CDT) approach (see [3] for a review), one constructs the state sum as

$$Z_{\text{CDT}} = \sum_{T \in \mathcal{T}} e^{iS_R(a,b)/l_P^2}, \tag{6.1}$$

where the sum goes over a class of triangulations $\mathcal{T}$ specified by causality requirements, while all 4-simplices in $T$ are isosceles, i.e. labeled by edge lengths $a$ and $b$, such that one can distinguish foliations of $T$ into space-like hypersurfaces labeled exclusively by edge lengths $a$, while each two hypersurfaces are connected by edge lengths $b$, which is usually chosen to be time-like rather than space-like (although this is not mandatory). This means that all tetrahedra within a given hypersurface are equilateral, while all 4-simplices filling up a slice of spacetime between two hypersurfaces are isosceles. In particular, of all possible isosceles 4-simplices (for given $a, b$ there are in total 40 inequivalent ones up to reflections and rotations, see Appendix F), exactly two are used, depicted by the following two diagrams:

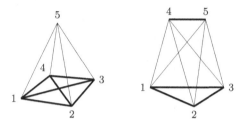

In both diagrams, thick lines are of length $a$, while thin lines are of length $b$. In the notation of [3] these two 4-simplices are denoted as $(4, 1)$ and $(3, 2)$ respectively.

One of the results obtained in section 3.5.5 (see also [125]) was the state sum (3.96), which is readily comparable to (6.1). The main difference, however, lies in the fact that for the generic choice of $\bar{m}$'s in (3.96) all 4-simplices in the triangulation must be identical, as dictated by the simplicity constraint. This is in contrast to (6.1), where two types of 4-simplices are being used. However, it turns out that this case is also covered by (3.96), as we will now show.

As was discussed in section 3.5.5, one can label the whole triangulation by labeling one arbitrary initial 4-simplex, and then employing the simplicity constraint to fix the labels for all other 4-simplices. In a generic case, this procedure is unique. However, if the labeling of the initial 4-simplex is chosen in a special way, there can be more freedom for the choice of other 4-simplices. As an example, choose the labels of the initial 4-simplex such that it is isosceles, for example the left one of the two diagrams above. If one assigns length $a$ to thick edges and length $b$ to thin edges, the simplicity constraint for that 4-simplex reduces to the following two equations:

$$\gamma l_p^2 |m_1| = A_H(a, a, a) \equiv \frac{a^2 \sqrt{3}}{4}, \quad \gamma l_p^2 |m_2| = A_H(a, b, b) \equiv \frac{a}{4} \sqrt{4b^2 - a^2}.$$
$$(6.2)$$

This is because only triangles of type $(a, a, a)$ and $(a, b, b)$ appear in the 4-simplex. Moreover, it is obvious that the simplicity constraint can always be uniquely solved for $a$ and $b$, given any choice of integers $m_1$ and $m_2$:

$$a = 2l_p \sqrt{\frac{\gamma |m_1|}{\sqrt{3}}}, \quad b = l_p \sqrt{\frac{\gamma |m_1|}{\sqrt{3}}} \sqrt{1 + 3\left(\frac{m_2}{m_1}\right)^2}.$$

Nevertheless, note that in this case there is *another* 4-simplex that is also made only of those two triangles, namely the right 4-simplex in the diagram above. Its simplicity constraint is identical to (6.2), and thus already satisfied. Therefore, in this case the 4-simplices in the triangulation need not be all identical, which leaves more flexibility in the possible labeling of the triangulation. In this case, restricting (3.96) to isosceles 4-simplices of the above type, we obtain

$$Z^{\text{sq}} = \sum_{m_1 \in \mathbb{Z}} \sum_{m_2 \in \mathbb{Z}} \left( \sum_{T \in \mathcal{T}} e^{i S_R(a,b)} \right), \tag{6.3}$$

where the expression in the parentheses is now precisely the CDT state sum (6.1). Restricting oneself to this choice of labeling, after "freezing out" the values for $m_1$ and $m_2$, the spin-cube model with the strongly imposed simplicity constraint reduces to the CDT model of quantum gravity.

It is worth noting that there are other isosceles 4-simplices for which the analogous construction may apply. In particular, there are two more pairs of 4-simplices which are labeled by only two edge lengths $a, b$ and consist of only two types of triangles (see Appendix F for details). One can choose those alternative geometries to satisfy the simplicity constraint in a similar way, and obtain a state sum similar to CDT. This is however not equivalent to CDT, because such 4-simplices do not induce a natural foliation of spacetime into space and time, which is one of the important aspects of CDT.

The presence of the CDT state sum (6.1) as one piece of the spin-cube state sum (3.96) essentially means that the spin-cube models contain the CDT model as a special case. Specifically, if one studies only the isosceles configurations of 4-simplices within a spin-cube model, one will recover all the wealth of results that can be obtained within the CDT approach to quantum gravity. In addition, spin-cube models allow one to study non-CDT-like isosceles configurations mentioned above, as well as non-isosceles configurations, all of which can potentially give rise to novel effects, not present in the CDT approach. Extending quantum gravity models from the CDT state sum to the spin-cube state sum is therefore useful, both as a way to study more general spacetime configurations, and as a way to study the relationship between the CDT formalism and the spin foam formalism, which is a very interesting open problem.

### 6.2.1 *Topological restrictions*

In the context of spin-cube models of QG, so far we have not discussed the topological and combinatorial properties of the manifold $M_4$ and its triangulation $T(M_4)$. There are two main questions to be addressed:

(Q1) Does the form of the state sum (3.96) place any restrictions on the choice of the topology of the manifold $M_4$?

(Q2) Does the choice of the topology of $M_4$ place any restrictions on the form of (3.96)?

Before we even begin discussing these questions, two remarks are necessary. First, it should be noted that we are interested in the realistic choice of four spacetime dimensions, which means that we are discussing 4-dimensional topological manifolds. The full classification of 4-dimensional manifolds is known to be an undecidable problem in the sense of Gödel's first incompleteness theorem [78, 109], and even among simply-connected 4-manifolds there are those which do not admit a triangulation to begin with, like the $E_8$ manifold [47, 117]. It is thus clear that any conclusive analysis of the above two questions is hopeless. This section is therefore devoted to discussing some examples and reformulations of the questions, without giving any general answers.

Second, one should keep in mind that the topology of $M_4$ is kept fixed when summing over different triangulations in (3.96). Given one initial triangulation $T(M_4)$, one can construct other, different triangulations of the manifold with the same topology, using for example Pachner moves, see [103]. Some suitable subset of these triangulations will define the domain $\mathcal{T}$ for the sum over triangulations. Additionally summing over different topologies of $M_4$ would represent the formalism of third quantization,

$$Z^{\text{tq}} = \sum_{\text{topologies}} Z^{\text{sq}},$$

which is out of the scope of this book.

After these introductory remarks, we can concentrate on what can be said regarding the questions (Q1) and (Q2) above. To this end, it is instructive to rewrite the state sum (3.96) as

$$Z^{\text{sq}} = Z_{10} + \cdots + Z_1,$$

where

$$Z_{10} = \sum_{\substack{\bar{m}_1,\ldots,\bar{m}_{10} \\ \bar{m}_1 \neq \ldots \neq \bar{m}_{10}}} \sum_{\alpha} \left( \sum_{T \in \mathcal{T}} e^{iS_R(L_\alpha(\bar{m}))} \right),$$

$$Z_9 = \sum_{\substack{\bar{m}_1,\ldots,\bar{m}_9 \\ \bar{m}_1 \neq \ldots \neq \bar{m}_9}} \sum_{\alpha} \left( \sum_{T \in \mathcal{T}} e^{iS_R(L_\alpha(\bar{m}))} \right),$$

$$\vdots$$

$$Z_2 = \sum_{\substack{\bar{m}_1,\bar{m}_2 \\ \bar{m}_1 \neq \bar{m}_2}} \sum_{\alpha} \left( \sum_{T \in \mathcal{T}} e^{iS_R(L_\alpha(\bar{m}))} \right),$$

$$Z_1 = \sum_{\bar{m}_1} \sum_{\alpha} \left( \sum_{T \in \mathcal{T}} e^{iS_R(L_\alpha(\bar{m}))} \right).$$

Here we have rearranged the summation over the ten $\bar{m}$ integers such that we can explicitly distinguish the pieces of the state sum where two, three, etc. of them are equal. The $Z_{10}$ term corresponds to the generic situation when all $\bar{m}$'s are mutually different. The $Z_1$ term corresponds to the case where all of $\bar{m}$'s are equal. The $Z_2$ term contains the generalized CDT state sum (6.3). In relation to questions (Q1) and (Q2) above, it will turn out that the $Z_{10}$ term will be most useful to discuss (Q2), while the $Z_1$ term is the most useful to discuss (Q1).

The second ingredient we need is the 1-complex dual to the triangulation. It is constructed as follows: to each 4-simplex in the triangulation one assigns a vertex, and to each tetrahedron common to two 4-simplices one assigns a link connecting two vertices. If the tetrahedron is on the boundary, then it belongs to only one 4-simplex, and the corresponding link is open-ended. Thus the 1-complex dual to the triangulation is some 5-valent graph with links encoding the adjacency relation among 4-simplices, while open-ended links encode boundary tetrahedra of the triangulation.

Looking first at the generic case of $Z_{10}$, one can note that all 4-simplices must be identical, while having 5 different irregular tetrahedra on their boundaries. The 4-simplices are then glued to each other along the identical tetrahedra, keeping the simplicity constraint satisfied. However, there exists a possibility that, after a certain number of gluings, two distant parts of the triangulation are supposed to "meet" due to topological structure, and at the meeting point they may fail to have compatible tetrahedra for consistent gluing. In this way topology may invalidate the generic choice of $\bar{m}$'s in $Z_{10}$, and thus place a restriction on the form of the state sum $Z^{\mathrm{sq}}$ (see question (Q2)).

This problem is actually equivalent to the problem of consistently coloring a 1-complex dual to the triangulation such that each link is colored by a number from the set $\{1, \ldots, 5\}$, and such that at every vertex no two colors are repeated. There is no known general answer to this question. Indeed, there may be 1-complex graphs for which such a coloring is possible, and graphs for which it is impossible. One obvious example of the former is the construction where we start from one vertex, connect five new vertices to its five links, and then keep connecting new vertices to new free links. But we do this in such a way that we never mutually connect two already existing free links, i.e. we never make a loop. The 1-complex graph constructed in such a way can always be colored in the required way — one starts by arbitrarily coloring the links of the initial vertex, and expands the coloring from there, keeping the coloring of links entering into each new vertex consistent with the previous choices. For the opposite example, we do not have any obvious examples, but it is certainly plausible that for some graphs consistent coloring may be impossible.

The above analysis suggests that the topology of the manifold might exclude the generic piece $Z_{10}$ from the total state sum, but it might also accommodate it without problems. Regarding the question (Q2) above, this is the best possible answer one can give at this stage — depending on the actual choice of the topology of $M_4$, there might or might not be some restrictions on the state sum.

On the other extreme, let us look at $Z_1$. This piece of the state sum contains equilateral 4-simplices. In an equilateral 4-simplex, there is only one type of tetrahedron available, and any two 4-simplices can be glued along any of their tetrahedra. In the language of the dual 1-complex, this means that we are supposed to color the links of the graph with a single color (vertices are now such that all "five" colors must be mutually identical). It is pretty obvious that every graph can be colored with a single color, without restrictions. This means that a triangulation corresponding to arbitrary topology can always be labeled with equilateral 4-simplices.

The problem with this, however, is that equilateral 4-simplices can only be embedded into a spacetime of Euclidean signature, while we are interested in the realistic Lorentzian case. So the equilateral 4-simplices actually have to be excluded from the state sum on the grounds of desired spacetime signature. So the only available 4-simplices in $Z^{\text{sq}}$ are isosceles (belonging to $Z_2$), with at least three different types of distinct tetrahedra. In the language of the dual 1-complex, this means that we are supposed to label the links of the graph with three distinct colors $\{1, 2, 3\}$, along with certain combinatorial rules for the vertices. Whether or not this is possible for a graph corresponding to the triangulation of an arbitrary topology, is again a problem for which there is no general answer. Like in the generic case, one can certainly construct graphs such that this coloring is possible. In particular, every graph that can be consistently colored with five different colors (corresponding to the $Z_{10}$ piece of the state sum), can also be colored using only three distinct colors (simply by identifying the fourth and fifth color with one of the previous three). For the opposite case we again do not have an explicit example, but it might exist.

The analysis of the $Z_1$ case then suggests an answer to the question (Q1), albeit an inconclusive one — the state sum may place restrictions on the possible topologies of $M_4$, but also maybe does not, depending on the existence of the dual 1-complex which cannot be colored with three distinct colors.

Finally, all remaining pieces of the state sum, $Z_9$ to $Z_2$ are "in between" the two extreme cases $Z_{10}$ and $Z_1$, and add no new insight. As one goes from $Z_1$ to $Z_{10}$, each piece $Z_k$ for higher $k$ places more stringent rules for coloring the graph, ranging from three colors with repetitions at vertices to five colors without repetitions. In this sense their analysis does not add any novel information about possible topologies that is not already present in cases $Z_1$ and $Z_{10}$.

The inconclusiveness of the answers to questions (Q1) and (Q2) given by the qualitative analysis of this section might seem underwhelming. One should therefore keep in mind that the undecidability of the classification of topological 4-manifolds severely limits what one can say about any general question involving the interaction between the structure of the state sum and topology, in the following way. Regarding the question (Q2), the topology of $M_4$ places restrictions on the state sum if no triangulation of $M_4$ can accommodate the $Z_{10}$ piece of the state sum. If we denote as $S$ the set of all triangulations $T$ that have some chosen fixed topology of $M_4$,

$$S = \{T \mid T \text{ is homeomorphic to } M_4\},$$

then one can rewrite the statement that topology places restrictions on the state sum as follows:

$$\forall T \ T \in S \Rightarrow \left( \begin{array}{l} \text{dual graph of } T \text{ cannot be colored} \\ \text{in a way compatible with } Z_{10} \end{array} \right).$$

But this statement is impossible to prove, since it is undecidable whether any given triangulation has the topology of $M_4$, so the set of triangulations with appropriate topology, $S$, cannot be effectively quantified over. In other words, the antecedent in the implication is undecidable, which makes the whole sentence undecidable.

Regarding the question (Q1), the state sum places restrictions on the topology of $M_4$ if there is no triangulation of $M_4$ that can accommodate the $Z_1$ piece of the state sum:

$$\forall T \quad T \in S \Rightarrow \begin{pmatrix} \text{dual graph of } T \text{ cannot be colored} \\ \text{in a way compatible with } Z_1 \end{pmatrix}.$$

Again, this is impossible to prove, for the same reason as before — it is undecidable whether or not any given triangulation has the topology of $M_4$, so the set of candidate triangulations cannot be quantified over.

Nevertheless, the analysis presented above gives us some practical tools and techniques to study particular cases. For example, given some particular triangulation, one can always construct its dual 1-complex, and determine whether or not it can be colored according to the rules given above. This question is decidable for any finite graph, since there are finitely many possible choices for the color of each link. One can simply try out all possible combinations of colors for all available links in the graph, and the algorithm is guaranteed to halt after a finite number of steps, since there are finite number of possibilities to test.

## 6.3 PLQG and other discrete QG models

As we have commented in chapter 1, the other approaches to QG which are based on a PL manifold for a triangulation of a smooth manifold, like the SF models or CDT, take the PL structure as an auxiliary tool in order to define the QG theory on a smooth manifold. Also in the Regge formulation of the path integral, this is the most popular approach. As we have mentioned, the direct attempts to define the smooth limit of the path integral via $L_\epsilon \to 0$, $\epsilon = 1, 2, \ldots, E$, and $E \to \infty$, have failed. However, we have demonstrated that the smooth limit can be applied to the effective action in the form of the smooth manifold approximation, see section 4.6.

Note that there is a third strategy which can be used to solve the problem of the smooth limit for a PLQG theory, and it is based on the Wilson theory of critical phenomena, see [59].

### 6.3.1  *Wilsonian approach to QG*

In the Wilsonian approach, one writes the Regge path integral as

$$Z_R(\beta) = \int \prod_\epsilon dL_\epsilon\, \mu(L)\, e^{-\beta S_R(L)},$$

where $\beta = -i/G_N\hbar$, see [59]. In this form, $Z_R$ can be recognized as a statistical partition function for an imaginary temperature $T = 1/k_B\beta$, where $k_B$ is the Boltzmann constant. One then looks for a value of $\beta$ for which $Z_R$ is divergent, which is a sign of a phase transition. In the gravitational context, this phase transition is interpreted as a transition from the PL manifold phase into a smooth manifold phase, since at the critical point the correlation length diverges, and the correlation length is proportional to the inverse lattice spacing. The theory near the critical point is then described by a QFT on a smooth manifold.

The classic example is the two-dimensional (2d) Heisenberg model, which is a system of spins on a two-dimensional regular lattice. Near the critical point, the theory is described by a 2d conformal QFT. In the case of Regge GR, the problem is that the critical point corresponds to a small $|\beta|$, which means that $G_N$ is large, so that one cannot use the perturbative QFT methods. On the other hand, in the EA approach, $G_N$ is a constant, and the perturbation parameter is $l_P^2 = G_N\hbar$, which gives a very small characteristic length, so that the smooth manifold approximation is described by the perturbative GR QFT with a cutoff proportional to the inverse average edge length.

Another assumption of the Wilsonian approach is that the coupling constants "run", i.e. they are functions of a momentum scale, which can be understood as an inverse characteristic length scale. This idea is a reasonable assumption for a renormalizable QFT, but in the case of GR, we have a non-renormalizable QFT, and therefore it is not clear how to obtain a finite number of running coupling constants. Weinberg proposed the idea of the asymptotically safe QFT, which is a way to define a finite number of running coupling constants for non-renormalizable QFTs [128]. This proposal was later developed into asymptotically safe QG (ASQG), see [26,43]. The main assumption in the ASQG approach is the existence of a certain fixed point for the running of the infinite number of the coupling constants in the GR QFT. At such a fixed point only a finite number of the coupling constants will be relevant and one can have a predictive theory. This conjecture sounds reasonable, but the proof is still missing.

In the case of Regge PLQG, although $G_N$ is constant, one can define the running in the effective QFT theory which describes the smooth-manifold approximation. In this case $G_N \to G_N(K)$, where $K \propto 1/\bar{L}$ is the momentum cutoff defined by the average space-like edge length in the triangulation slab corresponding to the present universe, see section 5.3.1. Although GR as a QFT is not renormalizable, GR regarded as a QFT with a physical cutoff is perturbatively finite, and in a perturbatively finite QFT one can still define the running of the coupling constants, see [55]. Namely, in a perturbatively finite QFT, a renormalized coupling is a function of $K/m$, where $m$ is a variable mass scale, and $K$ is a parameter which makes the Feynman diagrams finite, see [55]. In the case of Regge PLQG this parameter is the physical cutoff $1/\bar{L}$.

### 6.3.2  *Casual Set Theory*

Causal Set Theory (CST) postulates that spacetime is a set with a causal order, where the elements of the set are interpreted as the events in spacetime, see [119] for a review and references. The embedding of this set into a smooth manifold defines the distances between the events and the corresponding metric. After the spacetime embedding, CST is equivalent to a Regge PLQG for a $T(M)$ where the vertices obey the causality relations. Hence in CST the basic objects are the points, which induce the edge lengths, while in Regge PLQG one starts from the edge-lengths as the basic DoF. Although CST assumes a pre-geometry, which may have some physical implications, CST effectively reduces to the Regge QG.

# Appendix A

# 2-groups

In higher category theory (see [8] for an introduction), one can study particular generalizations of the algebraic structure of a group, called a 2-group (and higher — 3-group, 4-group, and $n$-group in general). Within that framework, an ordinary group is a category with a single element, while all morphisms are invertible. Generalizing to the next level, a notion of a 2-group is, by definition, a 2-category with a single element, while all morphisms and 2-morphisms are invertible.

In physics, one can describe a gauge symmetry by using a 2-group, in particular a strict Lie 2-group, which has been shown to be isomorphic to an algebraic structure called a crossed module. While the notion of a 2-group is conceptually very clean, the notion of a crossed module is algebraically more concrete, and hence more useful for practical calculations in physics. To that end, we give a very brief self-contained definition of a crossed module, and its corresponding differential crossed module, which is the analog of a Lie algebra of a given Lie group. For further details, see for example [83, 126].

### Pre-crossed module and crossed module

By definition, a *pre-crossed module* $(H \xrightarrow{\partial} G, \rhd)$ of groups $G$ and $H$, is given by a group map $\partial : H \to G$, together with a left action $\rhd$ of $G$ on $H$, by automorphisms, such that for each $h_1, h_2 \in H$ and $g \in G$ the following identity holds:

$$g\partial(h)g^{-1} = \partial(g \rhd h).$$

In a pre-crossed module the *Peiffer commutator* is defined as:

$$\langle h_1, h_2 \rangle_{\mathrm{p}} = h_1 h_2 h_1^{-1} \partial(h_1) \rhd h_2^{-1}.$$

Then, a pre-crossed module is said to be a *crossed module* if all of its Peiffer commutators are trivial, which is to say that

$$(\partial h) \triangleright h' = h h' h^{-1},$$

i.e. the *Peiffer identity* is satisfied.

### Differential pre-crossed module, differential crossed module

By definition, a *differential pre-crossed module* $(\mathfrak{h} \xrightarrow{\partial} \mathfrak{g}, \triangleright)$ of algebras $\mathfrak{g}$ and $\mathfrak{h}$ is given by a Lie algebra map $\partial : \mathfrak{h} \to \mathfrak{g}$ together with an action $\triangleright$ of $\mathfrak{g}$ on $\mathfrak{h}$ such that for each $h \in \mathfrak{h}$ and $g \in \mathfrak{g}$:

$$\partial(g \triangleright h) = [g, \partial(h)].$$

The action $\triangleright$ of $\mathfrak{g}$ on $\mathfrak{h}$ is on left by derivations, i.e. for each $h_1, h_2 \in \mathfrak{h}$ and each $g \in \mathfrak{g}$:

$$g \triangleright [h_1, h_2] = [g \triangleright h_1, h_2] + [h_1, g \triangleright h_2].$$

In a differential pre-crossed module, the Peiffer commutators are defined for each $h_1, h_2 \in \mathfrak{h}$ as:

$$\langle h_1, h_2 \rangle_p = [h_1, h_2] - \partial(h_1) \triangleright h_2.$$

The Peiffer commutator map $(h_1, h_2) \in \mathfrak{h} \times \mathfrak{h} \to \langle h_1, h_2 \rangle_p \in \mathfrak{h}$ (also called the *Peiffer paring*) is a bilinear $\mathfrak{g}$-equivariant map, i.e. all $h_1, h_2 \in \mathfrak{h}$ and $g \in \mathfrak{g}$ satisfy the following identity:

$$g \triangleright \langle h_1, h_2 \rangle_p = \langle g \triangleright h_1, h_2 \rangle + \langle h_1, g \triangleright h_2 \rangle_p.$$

A differential pre-crossed module is said to be a *differential crossed module* if all of its Peiffer commutators vanish, which is to say that for each $h_1, h_2 \in \mathfrak{h}$:

$$\partial(h_1) \triangleright h_2 = [h_1, h_2].$$

# Appendix B

# Proof that $\beta^a = 0$

We start from the equation

$$e^{[a} \wedge \beta^{b]} = 0\,.$$

By working in a coordinate basis, one can rewrite this equation in the component form as

$$\varepsilon^{\mu\nu\rho\sigma}\left(e^a{}_\mu \beta^b{}_{\rho\sigma} - e^b{}_\mu \beta^a{}_{\rho\sigma}\right) = 0\,. \tag{B.1}$$

By assuming that $\det(e^a{}_\mu) \neq 0$, we denote the inverse tetrads as $e^\mu{}_a$. First we contract equation (B.1) with $\varepsilon_{\alpha\nu\gamma\delta}e^\alpha{}_a e^\lambda{}_b$ to obtain

$$\varepsilon_{\alpha\nu\gamma\delta}\varepsilon^{\alpha\nu\rho\sigma}\beta^\lambda{}_{\rho\sigma} - \varepsilon_{\alpha\nu\gamma\delta}\varepsilon^{\lambda\nu\rho\sigma}\beta^\alpha{}_{\rho\sigma} = 0\,,$$

where $\beta^\lambda{}_{\rho\sigma} \equiv e^\lambda{}_a \beta^a{}_{\rho\sigma}$. Next, we use the identities

$$\varepsilon^{\lambda\mu\nu\rho}\varepsilon_{\lambda\alpha\beta\gamma} = -\det\begin{bmatrix} \delta^\mu_\alpha & \delta^\mu_\beta & \delta^\mu_\gamma \\ \delta^\nu_\alpha & \delta^\nu_\beta & \delta^\nu_\gamma \\ \delta^\rho_\alpha & \delta^\rho_\beta & \delta^\rho_\gamma \end{bmatrix}\,, \quad \varepsilon^{\mu\nu\rho\sigma}\varepsilon_{\mu\nu\alpha\beta} = -2\det\begin{bmatrix} \delta^\rho_\alpha & \delta^\rho_\beta \\ \delta^\sigma_\alpha & \delta^\sigma_\beta \end{bmatrix}\,,$$

in order to eliminate the contractions of the Levi-Civita symbols. After some algebra, we obtain

$$\beta^\lambda{}_{\gamma\delta} + \beta^\sigma{}_{\gamma\sigma}\delta^\lambda_\delta - \beta^\sigma{}_{\delta\sigma}\delta^\lambda_\gamma = 0\,. \tag{B.2}$$

By contracting the indices $\lambda$ and $\delta$, we immediately obtain $\beta^\sigma{}_{\gamma\sigma} = 0$. Substituting this back into (B.2), it follows that

$$\beta^\lambda{}_{\gamma\delta} = 0\,.$$

Finally, contracting with the tetrad $e^a{}_\lambda$ one obtains the result $\beta^a = 0$.

# Appendix C

# Regge EA perturbative expansion

Consider the following integro-differential equation

$$e^{i\Gamma(L)/\varepsilon} = \int_{-L}^{\infty} dl \, e^{i[S(L+l)-\Gamma'(L)l]/\varepsilon} , \qquad (C.1)$$

where $S(L)$ is a $C^{\infty}$ function, $L > 0$ and $\varepsilon$ is a small parameter. We will also assume that

$$S(L) = S_0(L) - i\varepsilon \log \mu(L) ,$$

where $\mu(L)$ is a measure which makes the integral

$$\int_0^{\infty} \mu(L) e^{iS_0(L)/\varepsilon} dL$$

convergent.

We want to solve equation (C.1) perturbatively in $\varepsilon$ as

$$\Gamma(L) = S(L) + \sum_{n>0} \varepsilon^n \Gamma_n(L) ,$$

up to an additive constant. Since

$$S(L+l) = S(L) + \sum_{n>0} S_n(L) \, l^n ,$$

where $S_n(L) = S^{(n)}(L)/n!$, we obtain

$$\Gamma_1 + \varepsilon\Gamma_2 + \varepsilon^2\Gamma_3 + \cdots = (-i) \log \int_{-L}^{\infty} dl \exp\left[\frac{i}{\varepsilon} S_2 l^2 - i\bar{\Gamma}_1' l + \frac{i}{\varepsilon} \sum_{n>2} S_n l^n\right] ,$$

$$(C.2)$$

where $\bar{\Gamma}_1 = \Gamma_1 + \varepsilon\Gamma_2 + \varepsilon^2\Gamma_3 + \cdots$. This equation has a solution since the integral in (C.2) can be written as

$$I = \int_{-L}^{\infty} dl \, e^{-sl^2/\varepsilon + wl} \exp\left(\sum_{n>2} s_n \, l^n\right) ,$$

155

where $s = -iS_2$, $s_n = iS_n/\epsilon$ and $w = -i\bar{\Gamma}_1'$. By expanding the exponent in the Maclaurin series, we obtain

$$I = \int_{-L}^{\infty} dl\, e^{-sl^2/\epsilon + wl} \left( 1 + \sum_{n>2} \hat{s}_n l^n \right) ,$$

where $\hat{s}_n$ are products of $s_n$. We can write the integral $I$ as

$$I = I_0 + \sum_n I_n \hat{s}_n ,$$

where

$$I_n = \int_{-L}^{\infty} dl\, e^{-s/\epsilon l^2 + wl}\, l^n .$$

The integrals $I_n$ can be calculated by differentiating $I_0$ with respect to $w$. It is easy to show that

$$I_0 = \sqrt{\frac{\pi\epsilon}{4s}}\, e^{\frac{\epsilon w^2}{4s}} \left[ 1 + \mathrm{erf}\left( L\sqrt{\frac{s}{\epsilon}} + \frac{w\sqrt{\epsilon}}{2\sqrt{s}} \right) \right] ,$$

where

$$\mathrm{erf}(x) = \frac{2}{\sqrt{\pi}} \int_0^x e^{-t^2}\, dt .$$

The domain of the error function can be extended to any complex number $z$ by using the Maclaurin expansion

$$\mathrm{erf}(z) = \frac{2}{\sqrt{\pi}} \sum_{n=0}^{\infty} \frac{(-1)^n z^{2n+1}}{(2n+1)n!} .$$

For large $|z|$ we can use

$$\mathrm{erf}(z) = 1 + \frac{e^{-z^2}}{z\sqrt{\pi}} \left( 1 + \sum_{n=1}^{N-1} \frac{(-1)^n (2n-1)!!}{2^n z^{2n}} + R_N(z) \right) , \qquad (C.3)$$

where $R_N(z) = O(z^{-2N})$ and (C.3) is valid for $|arg(z)| < 3\pi/4$ [2]. When $|arg(z)| < \pi/2$ one has

$$R_N(z) = \frac{(-1)^N (2N-1)!!}{2^N z^{2N}} \theta(z) ,$$

where

$$\theta(z) = \int_0^{\infty} e^{-t} (1 + t/z^2)^{-N-1/2}\, dt ,$$

while for $|arg(z)| < \pi/4$ one has $|\theta| < 1$, see [2].

Given the above formulas, one can show that as $\varepsilon \to 0$ and $L \to \infty$

$$I_0 \approx \sqrt{\frac{\pi \varepsilon}{s}} \, e^{\frac{\varepsilon w^2}{4s}} \left( 1 + \frac{\sqrt{\varepsilon} e^{-L^2 s/\varepsilon}}{2L\sqrt{s\pi}} \, e^{Lw} \right) .$$

It is convenient to introduce $\tilde{I}_0 = I_0/\sqrt{\varepsilon}$, then

$$\tilde{I}_{2k-1} = O(\varepsilon^k) , \quad \tilde{I}_{2k} = O(\varepsilon^k) , \quad k = 1, 2, 3, \ldots .$$

Note that there will be terms in $\tilde{I}_n$ proportional to half-integer powers of $\varepsilon$ smaller than $n/2$. However, these terms will be always multiplied by the factor $e^{-L^2 s/\varepsilon + Lw}$, which will suppress these terms provided

$$Re \, s = -(\log \mu)'' > 0 ,$$

for $L \to \infty$.

The equation (C.2) can be then written for large $L$ as

$$\log \left( \tilde{I}_0(L) + \sum_{n \geqslant 1} \hat{I}_n(L) \varepsilon^n \right) = \Gamma_1(L) + \varepsilon \Gamma_2(L) + \varepsilon^2 \Gamma_3(L) + \cdots , \quad \text{(C.4)}$$

where $\tilde{I}_n = \hat{I}_n \varepsilon^n$, and we have discarded on the LIIS of (C.4) the constant $\log \sqrt{\varepsilon}$ as well as the other constants. By using

$$\log \left( \tilde{I}_0(L) + \sum_{n \geqslant 1} \hat{I}_n(L) \varepsilon^n \right) = \log \tilde{I}_0(L) + \log \left( 1 + \sum_{n \geqslant 1} \frac{\hat{I}_n(L)}{\tilde{I}_0(L)} \varepsilon^n \right)$$

we can obtain $\Gamma_i$ as functions of $L$ by using the power-series expansion of $\log(1 + x)$ for small $x$.

# Appendix D

# Gaussian sums

Consider a sum

$$S(a) = \sum_{n=0}^{\infty} e^{-an^2},$$

where $a > 0$. It was shown in [19] that as $a \to 0$

$$S(a) = \sqrt{\frac{\pi}{4a}} + \frac{1}{2} e^{-a/4} \left[ \frac{\sinh \sqrt{a}}{\sqrt{a}} - \sum_{n=0}^{N} c_n \, a^{n+1/2} \, H_{2n+1}(\sqrt{a}/2) \right]$$
$$+ O(a^{N+3/2}),$$

where

$$c_n = \frac{(2^{2n+1} - 1)B_{2n+2}}{2^{2n}(2n+2)!},$$

$B_n$ are Bernoulli numbers and $H_n(x)$ are Hermite polynomials. If $a < 1$, then in the limit $N \to \infty$ we obtain

$$S(a) = \sqrt{\frac{\pi}{4a}} + \frac{1}{2} e^{-a/4} \left[ \frac{\sinh \sqrt{a}}{\sqrt{a}} - \sum_{n=0}^{\infty} c_n \, a^{n+1/2} \, H_{2n+1}(\sqrt{a}/2) \right].$$

Let $R(a) = S(a) - \sqrt{\frac{\pi}{4a}}$. Then

$$R(a) = \sum_{n=0}^{\infty} r_n \, a^{n/2},$$

for $a < 1$. We can now define a complex function

$$R(z) = \sum_{n=0}^{\infty} r_n \, z^{n/2},$$

for $|z| < 1$. Consequently we can define

$$S(z) = \sqrt{\frac{\pi}{4z}} + R(z),$$

for $0 < |z| < 1$, so that

$$S(-ia) \approx \sqrt{\frac{i\pi}{4a}}$$

as $a \to 0$.

## Appendix E

# Higher-loop matter contributions to the cosmological constant

In this section we will prove the formula (4.159) for the matter contributions to CC. The matter contributions are given by the sum of $n$-loop one-particle-irreducible (1PI) QFT Feynman diagrams with no external legs and with a momentum cut-off $\hbar K$. This is because the $\phi$-independent terms in the effective action are determined by the non-zero EA diagrams such that the $\phi \to 0$ limit is taken in the propagator and the vertex functions. This leaves only the matter 1PI vacuum-energy diagrams.

Let $U(\phi)$ be given by (4.147). Then the contribution to $\Lambda_m$ of $O(\hbar^n)$ is given by the sum of $n$-loop 1PI vacuum diagrams, which we denote as $\delta_n \Lambda_m$. This sum can be represented as

$$\delta_n \Lambda_m = \langle \; \text{\scriptsize⬭⬭} \cdots \text{◯} \; \rangle_n + \langle \; \text{⬭} \cdots \text{⬭} \; \rangle_n$$

$$+ \langle \; \text{✲} \; \rangle_n + \langle \; \text{⬡} \; \rangle_n + \cdots , \tag{E.1}$$

where the chain graphs appear for $n \geqslant 2$, watermelon graphs appear for $n \geqslant 3$, flower and polygon-in-a-circle graphs appear for $n \geqslant 4$, and so on.

We would like to determine the large-$K$ behavior of these graphs. This asymptotics is generically given by $O(K^D)$, where $D$ is the superficial degree of divergence of the graph. However, there are exceptions, and we will show that this happens in the case of flower graphs.

The 2-loop matter contribution to CC is given by the chain graph

$$\delta_2 \Lambda_m = c_2 \, \lambda \, l_P^4 \left( \int_0^K \frac{k^3 dk}{k^2 + \omega^2} \right)^2 \approx c_2 \, \lambda \, l_P^4 \, K^4 = c_2 \, \frac{l_P^4}{L_\lambda^2 L_K^4} \tag{E.2}$$

since $K \gg \omega$. This agrees with $D = 4$ for the 2-loop chain graph.

At 3 loops we have the chain graph contribution

$$\delta_3^C \Lambda_m = c_3 \, \lambda^2 \, l_P^6 \left( \int_0^K \frac{k^3 dk}{k^2 + \omega^2} \right)^2 \int_0^K \frac{q^3 dq}{(q^2 + \omega^2)^2} \approx c_3 \, \lambda^2 \, l_P^6 \, K^4 \ln(K^2/\omega^2) \,.$$
$$\tag{E.3}$$

This graph has $D = 4$ and the asymptotics (E.3) is consistent with this value of $D$.

For the 3-loop watermelon graph we obtain

$$\delta_3^M \Lambda_m = m_3 \, \lambda^2 \, l_P^6 \int_0^K \frac{k^3 dk}{k^2 + \omega^2} \int_0^K \frac{q^3 dq}{q^2 + \omega^2}$$
$$\int_{r \leqslant K} \frac{d^4 \vec{r}}{(r^2 + \omega^2)[(\vec{r} - \vec{k} - \vec{q})^2 + \omega^2]} \tag{E.4}$$
$$\approx m_3 \, \lambda^2 \, l_P^6 \, K^4 \ln(K^2/\omega^2) \,,$$

which again agrees with the corresponding $D$.

At 4 loops the flower graph appears, and it gives

$$\delta_4^F \Lambda_m = f_3 \, \lambda^3 \, l_P^6 \left( \int_0^K \frac{k^3 dk}{k^2 + \omega^2} \right)^3 \int_0^K \frac{q^3 dq}{(q^2 + \omega^2)^6} \,. \tag{E.5}$$

This integral has $D = 4$, but its asymptotics is given by $D = 6$. The reason is that the second integral is not asymptotic to $K^{-2}$ but it is asymptotic to a non-zero constant, so that

$$\delta_4^F \Lambda_m \approx f_4 \, l_P^2 K^4 \, \bar{\lambda}^3 (K/\omega)^2 \,. \tag{E.6}$$

An $n \geqslant 3$ chain graph gives

$$\delta_n^C \Lambda_\phi = c_n \, \lambda^{n-1} \, l_P^{2n} \left( \int_0^K \frac{k^3 dk}{k^2 + \omega^2} \right)^2 \left( \int_0^K \frac{k^3 dk}{(k^2 + \omega^2)^2} \right)^{n-2}$$

$$\approx c_n \, \lambda^{n-1} \, l_P^{2n} \, K^4 \left( \ln(K^2/\omega^2) \right)^{n-2} \,, \tag{E.7}$$

while an $n \geqslant 4$ polygon graph gives

$$\delta_n^P \Lambda_\phi = p_n \, \lambda^{n-1} \, l_P^{2n} \int_0^K \frac{k^3 dk}{k^2 + \omega^2} \int_0^K \frac{q^3 dq}{q^2 + \omega^2}$$
$$\left( \int_{r \leqslant K} \frac{d^4 \vec{r}}{(r^2 + \omega^2)[(\vec{r} - \vec{k} - \vec{q})^2 + \omega^2]} \right)^{n-2} \tag{E.8}$$
$$\approx p_n \, \lambda^{n-1} \, l_P^{2n} \, K^4 \left( \ln(K^2/\omega) \right)^{n-2} \,.$$

A flower graph gives for $n \geqslant 4$

$$\delta_n^F \Lambda_m \approx f_n \, l_P^2 K^4 \, \bar{\lambda}^{n-1} (K^2/\omega^2)^{n-3} \,. \tag{E.9}$$

As far as the other 1PI vacuum graphs are concerned, their $D$ is less than 4, and consequently the main contribution for large $K$ is given by

$$\Lambda_m \approx l_P^2 \, K^4 \left[ c_1 \ln(K^2/\omega^2) + \sum_{n \geqslant 2} c_n \bar{\lambda}^{n-1} \left( \ln(K^2/\omega^2) \right)^{n-2} \right. \\ \left. + \sum_{n \geqslant 4} d_n \bar{\lambda}^{n-1} \left( K^2/\omega^2 \right)^{n-3} \right], \tag{E.10}$$

where $\bar{\lambda} = \lambda \, l_P^2$ is dimensionless. Since $K \gg \omega$, we get

$$\Lambda_m \approx l_P^2 \, K^4 \sum_{n \geqslant 4} d_n \bar{\lambda}^{n-1} \left( K^2/\omega^2 \right)^{n-3} \,, \tag{E.11}$$

so that the flower graphs have a dominant contribution.

This expansion will be perturbative if

$$\bar{\lambda} K^2/\omega^2 < 1 \,. \tag{F.12}$$

Since $\bar{\lambda} = 1/8$ we obtain $K/\omega < \sqrt{8}$. On the other hand, the cutoff $K$ has to satisfy

$$K \gg \omega \,, \tag{E.13}$$

so that the series (E.11) will not be a perturbative series. Therefore we have to include the diagrams with a large number of loops in order to obtain an accurate value for $\Lambda_m$.

# Appendix F

# Isosceles 4-simplices

We define a given 4-simplex to be called "isosceles" iff all its edges have length $a$ or $b$, where by convention $a, b \in \mathbb{R}^+$. All triangles in such a simplex are either isosceles or equilateral. Also by convention, we exclude fully equilateral 4-simplices, by requiring that edges of both lengths $a$ and $b$ must be present in the 4-simplex.

It is an interesting question to determine all possible inequivalent isosceles 4-simplices. There are in total 20 different ones, up to the $a \leftrightarrow b$ exchange symmetry and permutations of the vertices. If we label the vertices of a 4-simplex as per the diagram below,

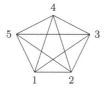

and if we order the 10 edge lengths into a 10-tuple as

$$(l_{12}, l_{13}, l_{14}, l_{15}, l_{23}, l_{24}, l_{25}, l_{34}, l_{35}, l_{45}),$$

the distinct labelings for the 20 isosceles 4-simplices are given as:

| | | | |
|---|---|---|---|
| 1: | $(a,a,a,a,a,a,a,a,a,b)$, | 11: | $(a,a,a,b,a,a,b,b,a,b)$, |
| 2: | $(a,a,a,a,a,a,a,a,b,b)$, | 12: | $(a,a,a,b,b,b,a,b,a,a)$, |
| 3: | $(a,a,a,a,a,a,b,b,a,a)$, | 13: | $(a,a,a,b,a,b,a,b,b,a)$, |
| 4: | $(a,a,a,a,a,a,b,a,b,b)$, | 14: | $(a,a,a,a,a,a,b,b,b,b)$, |
| 5: | $(a,a,a,b,a,a,b,b,a,a)$, | 15: | $(a,a,b,b,b,a,b,b,a,a)$, |
| 6: | $(a,a,a,a,a,a,b,b,a,b)$, | 16: | $(a,a,a,b,a,b,b,b,b,a)$, |
| 7: | $(a,a,a,a,a,a,a,b,b,b)$, | 17: | $(a,a,a,b,a,a,b,b,b,b)$, |
| 8: | $(b,b,a,a,a,b,a,a,a,a)$, | 18: | $(a,a,a,b,a,b,a,b,b,b)$, |
| 9: | $(a,a,a,b,a,a,b,a,b,b)$, | 19: | $(a,a,a,a,a,b,b,b,b,b)$, |
| 10: | $(a,a,a,b,a,a,b,b,a,b)$, | 20: | $(a,a,a,b,b,b,a,b,a,b)$. |

The above table has been computer-generated by brute force counting of all possibilities. If we mark all $a$-edges with thin lines and all $b$-edges with thick lines, these 4-simplices can be drawn as follows, respectively:

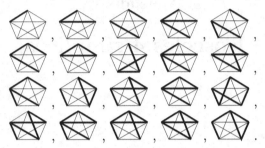

All the above 4-simplices have different 4-volumes, tetrahedra volumes and triangle areas. To these, one should also add 20 more 4-simplices, which are obtained by exchanging $a \leftrightarrow b$, i.e. by switching thick and thin lines. This gives a total of 40 inequivalent isosceles 4-simplices.

There are four different types of triangles that make up isosceles 4-simplices. These are $(a, a, a)$, $(a, a, b)$, $(a, b, b)$ and $(b, b, b)$ triangles. Most of the 4-simplices contain either three or sometimes all four types of triangles, making them unsuitable for satisfying the simplicity constraint, as explained in the main text. Nevertheless, there are five pairs of 4-simplices which are made up only of two types of triangles. These are:

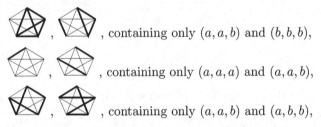

and their $a \leftrightarrow b$ duals (the third pair is self-dual). Of these pairs, one can use the first pair (or their dual) to construct a triangulation which features 3-dimensional hypersurfaces made of equilateral tetrahedra, such that these hypersurfaces are separated by "equal distances" everywhere. Thus, one can establish the triangulation-induced foliation of spacetime into space and time. This property is exploited by the CDT approach to quantum gravity. The remaining pairs above lack this property, because they mix edges of size $a$ and $b$ in a way that does not admit a nice $3 + 1$ foliation.

# Appendix G

# Fourier integral of a PL function

Consider a PL function $f(x)$ on an interval $[-NL/2, NL/2]$ such that the function takes constant values on $N$ subintervals of length $L$. This is the one-dimensional analog of the situation when a scalar field is described by a finite number of values at the dual vertices of a triangulation, so that $N$ can be interpreted as the total number of the cells and $L$ as the average edge length.

Let us extend $f(x)$ to a periodic function on $\mathbb{R}$ by $f(x + LN) = f(x)$. The corresponding Fourier series expansion is given by

$$f(x) = \frac{1}{\sqrt{NL}} \sum_{n=-\infty}^{\infty} c_n \, e^{\frac{2i\pi n}{NL} x} \,,$$

where

$$c_n = \frac{1}{\sqrt{NL}} \int_{-\frac{NL}{2}}^{\frac{NL}{2}} f(x) \, e^{-\frac{2i\pi n}{NL} x} \, dx \,.$$

On the other hand, the function

$$\tilde{f}(x) = \begin{cases} f(x), & |x| \leqslant NL/2 \\ 0, & |x| > NL/2 \end{cases}$$

can be written as a Fourier integral

$$\tilde{f}(x) = \frac{1}{\sqrt{2\pi}} \int_{-\infty}^{\infty} \tilde{\phi}(k) e^{ikx} \, dk \,,$$

where

$$\tilde{\phi}(k) = \frac{1}{\sqrt{2\pi}} \int_{-\infty}^{\infty} \tilde{f}(x) e^{-ikx} \, dx = \frac{1}{\sqrt{2\pi}} \int_{-\frac{NL}{2}}^{\frac{NL}{2}} f(x) e^{-ikx} \, dx \,.$$

Let us make an approximation in the Fourier series expansion

$$\sum_{n=-\infty}^{\infty} c_n\, e^{\frac{2i\pi n}{NL}x} \approx \sum_{n=-N}^{N} c_n e^{\frac{2i\pi n}{NL}x},$$

which can be justified for large $N$, i.e. when $N \to \infty$. In this case we can make another approximation

$$\sum_{n=-N}^{N} c_n\, e^{\frac{2i\pi n}{NL}x} \approx \int_{-N}^{N} \tilde{c}(u)\, e^{\frac{2i\pi u}{NL}x} du,$$

where we have replaced the discrete coefficients $c_n$ by a smooth function $\tilde{c}(u)$ on $[-N, N]$ such that $\tilde{c}(u = n) = c_n$. By introducing a new variable $k = 2u\pi/NL$, we obtain

$$f(x) \approx \frac{1}{\sqrt{NL}} \int_{-\frac{2\pi}{L}}^{\frac{2\pi}{L}} \tilde{c}(k) e^{ikx} dk. \tag{G.1}$$

Note that in the cutoff regularization in QFT we replace the infinite interval in the Fourier integral with a finite interval, so that

$$f(x) \to f_K(x) = \frac{1}{\sqrt{2\pi}} \int_{-K}^{K} \tilde{\phi}(k) e^{ikx}\, dk. \tag{G.2}$$

When (G.2) is compared to (G.1) we conclude

$$\tilde{\phi}(k) \approx \frac{\sqrt{2\pi}}{\sqrt{NL}} \tilde{c}(k)$$

for $|k| \leqslant 2\pi/L$ and

$$K = \frac{2\pi}{L},$$

i.e. the momentum cutoff is proportional to the inverse of the average edge length in a triangulation.

# Bibliography

[1] L. F. Abbott, *Acta Phys. Pol.* **B13**, 33-50 (1982).

[2] M. Abramowitz and I. A. Stegun, *Handbook of Mathematical Functions with Formulas, Graphs, and Mathematical Tables*, Dover Publications, New York (1972).

[3] J. Ambjorn, A. Görlich, J. Jurkiewicz and R. Loll, *Phys. Rep.* **519**, 127-210 (2012).

[4] A. Ashtekar, *Phys. Rev. Lett.* **57**, 2244 (1986).

[5] M. F. Atiyah, *Pub. Math. de l'Institut des Hautes Scientifiques* **68**, 175-186 (1988).

[6] J. C. Baez, *Lect. Notes Phys.* **543**, 25 (2000).

[7] J. C. Baez, A. Baratin, L. Freidel and D. K. Wise, *Mem. Amer. Math. Soc.* **219**, 1032 (2012).

[8] J. C. Baez and J. Huerta, *Gen. Relativ. Gravit.* **43**, 2335 (2011).

[9] J. C. Baez and D. K. Wise, *Commun. Math. Phys.* **333**, 153-186 (2015).

[10] A. Baratin and L. Friedel, *Class. Quant. Grav.* **24**, 2027 (2007).

[11] A. Baratin and D. K. Wise, *AIP Conf. Proc.* **1196**, 28-35 (2009).

[12] J. F. Barbero G., *Phys. Rev. D* **51**, 5507 (1995).

[13] J. Barrett, R. Dowdall, W. Fairbairn, F. Hellmann and R. Pereira, *Class. Quant. Grav.* **27**, 165009 (2010).

[14] J. W. Barrett and T. J. Foxon, *Class. Quant. Grav.* **11**, 543-556 (1994).

[15] J. W. Barrett and I. Naish-Guzman, *Class. Quant. Grav.* **26**, 155014 (2009).

[16] J. W. Barrett, M. Rocek and R. M. Williams, *Class. Quant. Grav.* **16**, 1373-1376 (1999).

[17] A. O. Barvinsky and G. A. Vilkovisky, *Nucl. Phys.* **B333**, 471-511 (1990).

[18] Z. Bern, J. Carrasco, W. Chen, A. Edison, H. Johansson, J. Parra-Martinez, R. Roiban and M. Zeng, *Phys. Rev. D* **98**, 086021 (2018).

[19] B. C. Berndt and B. Kim, *Proc. Amer. Math. Soc.* **139**, 3779 (2011).

[20] E. Bianchi, D. Regoli and C. Rovelli, *Class. Quant. Grav.* **27**, 185009 (2010).

[21] D. Birmingham, M. Blau, M. Rakowski and G. Thompson, *Phys. Rep.* **209**, 129 (1991).

[22] N. D. Birrell and P. C. W. Davies, *Quantum Fields in Curved Space*, Cambridge University Press, Cambridge (1982).

[23] M. Blau and G. Thompson, *Phys. Lett. B* **255**, 535-542 (1991).

[24] M. Blau and G. Thompson, *Ann. Phys.* **205**, 130-172 (1991).

[25] D. Bohm, *Phys. Rev.* **85**, 166 (1952).

[26] A. Bonanno, A. Eichhorn, H. Gies, J. M. Pawlowski, R. Percacci, M. Reuter, F. Saueressig and G. P. Vacca, *Front. Phys.* **8**, 269 (2020).

[27] R. Bousso and J. Polchinski, *JHEP* **6**, 006 (2000).

[28] M. Caicedo, R. Gianvittorio, A. Restuccia and J. Stephany, *Phys. Lett. B* **354**, 292 (1995).

[29] A. S. Cattaneo, P. Cotta-Ramusino and C. A. Rossi, *Lett. Math. Phys.* **51**, 301 (2000).

[30] A. Cayley, *Cambridge Math. Jour.* **2**, 267 (1841); K. Menger, *Math. Ann.* **100**, 75 (1928).

[31] M. Celada, D. González and M. Montesinos, *Class. Quant. Grav.* **33**, 213001 (2016).

[32] D. W. Chiou, *Int. J. Mod. Phys. D* **24**, 1530005 (2015).

[33] A. Connes, *Noncommutative Geometry*, Academic Press, San Diego (1994).

[34] F. Conrady and L. Freidel, *Class. Quant. Grav.* **25**, 245010 (2008).

[35] N. Costa-Dias, A. Miković and J. Prata, *J. Math. Phys.* **47**, 082101 (2006).

[36] L. Crane, L. H. Kauffman and D. N. Yetter, *Jour. Knot Theor. Ramif.* **06**, 177-234 (1997).

[37] L. Crane and M. D. Sheppeard, "2-Categorical Poincaré representations and state sum applications", arXiv:math/0306440.

[38] V. Cuesta and M. Montesinos, *Phys. Rev. D* **76**, 104004 (2007).

[39] V. Cuesta, M. Montesinos, M. Velázquez and J. D. Vergara, *Phys. Rev. D* **78**, 064046 (2008).

[40] G. D'Amico, N. Kaloper, A. Padilla, et al. *JHEP* **9**, 74 (2017).

[41] B. S. DeWitt, *Phys. Rev.* **160**, 1113-1148 (1967).

[42] B. Dittrich and S. Speziale, *New J. Phys.* **10**, 083006 (2008).

[43] A. Eichhorn, "Asymptotically safe gravity", *In search for the unexpected*, proceedings of the 57th Erice International School of Subnuclear Physics (2019).

[44] J. Engle, E. R. Livine, R. Pereira and C. Rovelli, *Nucl. Phys.* **B799**, 136-149 (2008).

[45] J. Engle, I. Vilensky and A. Zipfel, *Phys. Rev. D* **94**, 064025 (2016).

[46] J. Feldbrugge, J.-L. Lehners and N. Turok, *Phys. Rev. D* **95**, 103508 (2017).

[47] M. H. Freedman, *J. Differential Geom.* **17**, 357 (1982).

[48] D. Z. Freedman, P. van Nieuwenhuisen and S. Ferrara, *Phys. Rev. D* **13**, 3214 (1976).

[49] L. Freidel and K. Krasnov, *Class. Quant. Grav.* **25**, 125018 (2008).

[50] L. Freidel, D. Minić and T. Takeuchi, *Phys. Rev. D* **72**, 104002 (2005).

[51] L. Freidel and S. Speziale, *Phys. Rev. D* **82**, 084040 (2010).

[52] L. Freidel and J. Ziprick, *Class. Quant. Grav.* **31**, 045007 (2014).

[53] F. Girelli, H. Pfeiffer and E. M. Popescu, *J. Math. Phys.* **49**, 032503 (2008).

[54] M. Goroff and J. H. Schwarz, *Phys. Lett. B* **127**, 61-64 (1983).

[55] M. A. Green and J. W. Moffat, *Eur. Phys. Jour. Plus*, **136**, 919 (2021).

[56] M. B. Green, J. H. Schwarz and E. Witten, *Superstring Theory*, Cambridge University Press, Cambridge (1987).

[57] J. J. Halliwell and J. B. Hartle, *Phys. Rev. D* **41**, 1815 (1990).

[58] J. J. Halliwell and J. Louko, *Phys. Rev. D* **39**, 2206 (1989).

[59] H. W. Hamber, *Gen. Rel. Grav.* **41**, 817-876 (2009).

[60] J. B. Hartle and S. W. Hawking, *Phys. Rev. D* **28**, 2960-2975 (1983).

[61] P. R. Holland, *The Quantum Theory of Motion: An account of the De Broglie-Bohm causal interpretation of quantum mechanics*, Cambridge University Press, Cambridge (1993).

[62] S. Holst, *Phys. Rev. D* **53**, 5966 (1996).

[63] G. T. Horowitz, *Commun. Math. Phys.* **125**, 417 (1989).

[64] G. Immirzi, *Class. Quant. Grav.* **14**, L177-L181 (1997).

[65] C. Isham, *NATO Sci. Ser. C* **409**, 157 (1993).

[66] C. Isham, *Lect. Notes Phys.* **434** 1-21 (1994).

[67] C. Isham, *Lect. Notes Phys.* **434**, 150-169 (1994).

[68] S. Kachru, R. Kalosh, A. Linde and S. Trivedi, *Phys. Rev. D* **68**, 046005 (2003).

[69] H. Kleinert, *Path Integrals in Quantum Mechanics, Statistics, Polymer Physics and Financial Markets*, World Scientific (2006).

[70] H. Kodama, *Prog. Theor. Phys.* **80**, 1024 (1988).

[71] H. Kodama, *Int. J. Mod. Phys. D* **1**, 439 (1992).

[72] E. R. Livine and S. Speziale, *Phys. Rev. D* **76**, 084028 (2007).

[73] S. MacDowell and F. Mansouri, *Phys. Rev. Lett.* **38**, 739 (1977); Erratum, *Phys. Rev. Lett.* **38**, 1376 (1977).

[74] J. Madore, *An Introduction to Noncommutative Differential Geometry and its Physical Applications*, Cambridge University Press, Cambridge (1999).

[75] J. Makela, *Class. Quant. Grav.* **17**, 4991-4998 (2000).

[76] S. Mandelstam, *Phys. Lett. B* **277**, 82 (1992).

[77] M. Marino, *Chern-Simons Theory, Matrix Models, And Topological Strings*, Oxford University Press, Oxford (2005).

[78] A. Markov, *Dokl. Akad. Nauk SSSR*, **121**, 218 (1958).

[79] J. Martin, *Compt. Rend. Phys.* **13**, 566-665 (2012).

[80] J. F. Martins and A. Miković, *Commun. Math. Phys.* **279**, 381-399 (2008).

[81] J. F. Martins and A. Miković, *Commun. Math. Phys.* **288**, 745-772 (2009).

[82] J. F. Martins and A. Miković, *Adv. Theor. Math. Phys.* **15**, 1059 (2011).

[83] J. F. Martins and R. Picken, *Differ. Geom. Appl. Jour.* **29**, 179 (2011).

[84] A. Miković, *Class. Quant. Grav.* **19**, 2335-2354 (2002).

[85] A. Miković, *Class. Quant. Grav.* **21**, 3909-3922 (2004). Errata: *Class. Quant. Grav.* **23**, 5459 (2006).

[86] A. Miković, *Fortsch. Phys.* **56** (2008) 475-479.

[87] A. Miković, *Rev. Math. Phys.* **25**, 1343008 (2013).

[88] A. Miković, *Adv. Theor. Math. Phys.* **21**, 631 (2017).

[89] A. Miković and M. A. Oliveira, *Gen. Relativ. Gravit.* **47**, 58 (2015).

[90] A. Miković, M. A. Oliveira and M. Vojinović, *Class. Quant. Grav.* **33**, 065007 (2016).

[91] A. Miković, M. A. Oliveira and M. Vojinović, "Hamiltonian analysis of the BFCG theory for a strict Lie 2-group", accepted for publication in *Adv. Theor. Math. Phys.*, arXiv:1610.09621.

[92] A. Miković and M. Vojinović, *Adv. Theor. Math. Phys.* **15**, 801 (2011).

[93] A. Miković and M. Vojinović, "Graviton propagator asymptotics and the classical limit of ELPR/FK spin foam models", arXiv:1103.1428.

[94] A. Miković and M. Vojinović, *Class. Quant. Grav.* **28**, 225004 (2011).

[95] A. Miković and M. Vojinović, *Class. Quant. Grav.* **29**, 165003 (2012).

[96] A. Miković and M. Vojinović, *Jour. Phys. Conf. Ser.* **360**, 012049 (2012).

[97] A. Miković and M. Vojinović, *Class. Quant. Grav.* **30**, 035001 (2013).

[98] A. Miković and M. Vojinović, *Jour. Phys. Conf. Ser.* **532**, 012020 (2014).

[99] A. Miković and M. Vojinović, *Europhys. Lett.* **110**, 40008 (2015).

[100] A. Miković and M. Vojinović, *SFIN* **31**, 267 (2018).

[101] V. P. Nair, *Quantum Field Theory - A Modern Perspective*, Springer (2005).

[102] H. Ooguri, *Mod. Phys. Lett. A* **07**, 2799-2810 (1992).

[103] U. Pachner, *Eur. Jou. Combinat.* **12**, 129-145, (1991).

[104] A. Perez, *Living Rev. Rel.* **16**, 3 (2013).

[105] A. Perez and C. Rovelli, *Phys. Rev. D* **73**, 044013 (2006).

[106] J. F. Plebanski, *J. Math. Phys.* **12**, 2511 (1977).

[107] J. Polchinski, "The cosmological constant and the string landscape", *The Quantum Structure of Space and Time: Proceedings of the 23rd Solvay Conference on Physics*. Brussels, Belgium. 1-3 December 2005, D. Gross, M. Nenneaux and A. Servin (eds.) (2007).

[108] G. Ponzano and T. Regge, *Spectroscopic and Group Theoretical Methods in Physics: Racah Memorial Volume*, 75-103 (1968).

[109] B. Poonen, "Undecidable problems: a sampler", in *Interpreting Gödel: Critical essays*, ed. J. Kennedy, 211-241, Cambridge Univ. Press (2014).

[110] P. Ramond, *Field Theory: A Modern Primer*, Addison-Wesley Publishing Company Inc., Boston (1989).

[111] T. Regge, *Nuovo Cimento* **19**, 558-571 (1961).

[112] T. Regge and R. M. Williams, *J. Math. Phys.* **41**, 3964 (2000).

[113] C. Rovelli, *Quantum Gravity*, Cambridge University Press, Cambridge (2004).

[114] C. Rovelli, *Phys. Rev. Lett.* **97** (2006) 151301.

[115] W. Ruhl, *The Lorentz group and harmonic analysis*, WA Benjamin Inc, New York (1970).

[116] A. S. Schwarz, *Lett. Math. Phys.* **2**, 247 (1977).

[117] A. Scorpan, *The Wild World of 4-manifolds*, American Mathematical Society, Rhode Island (2005).

[118] S. Speziale and W. M. Wieland, *Phys. Rev. D* **86**, 124023 (2012).

[119] S. Surya, *Living Rev. Rel.* **22**, 5 (2019).

[120] L. Susskind, "The anthropic landscape of string theory", *Universe or Multiverse?*, B. Carr (ed.) Cambridge University Press, 247-266 (2007).

[121] V. Turaev and O. Viro, *Topology* **31**, 865 (1992).

[122] A. Vilenkin and M. Yamada, *Phys. Rev. D* **99**, 066010 (2019).

[123] M. Vojinović, *SFIN* **26**, 361 (2013).

[124]  M. Vojinović, *Gen. Relativ. Gravit.* **46**, 1616 (2014).
[125]  M. Vojinović, *Phys. Rev. D* **94**, 024058 (2016).
[126]  W. Wang, *Jour. Math. Phys.* **55**, 043506 (2014).
[127]  S. Weinberg, *Rev. Mod. Phys.* **61** 1 (1989).
[128]  S. Weinberg, *PoS CD* **09** 001 (2009).
[129]  C. Weinwright and R. Williams, *Class. Quant. Grav.* **21**, 4865-4880 (2004).
[130]  E. Witten, *Comm. Math. Phys.* **117**, 353-386 (1988).

Printed in the United States
by Baker & Taylor Publisher Services